S.T.(P)
Technology
Today Series

A Series for Technicians

MATHEMATICS
FOR TECHNICIANS

Level III
Engineering
Mathematics

S.T.(P)
Technology
Today Series

A Series for Technicians

MATHEMATICS FOR TECHNICIANS

Level III Engineering Mathematics

A. Greer
C.Eng., M.R.Ae.S.
Senior Lecturer
City of Gloucester College of Technology

G. W. Taylor
B.Sc.(Eng.), C.Eng., M.I.Mech.E.
Principal Lecturer
City of Gloucester College of Technology

Stanley Thornes (Publishers) Ltd.

First published in 1978 by:
Stanley Thornes (Publishers) Ltd
EDUCA House
32 Malmesbury Road
Kingsditch
CHELTENHAM GL51 9PL
England

ISBN 0 85950 0691

Typeset in Monotype 10/12 Times New Roman by
Gloucester Typesetting Co Ltd
Printed and bound in Great Britain by
The Pitman Press, Bath

CONTENTS

AUTHORS' NOTE ON SERIES   ix

NOTE ON THIS VOLUME   x

Chapter 1   **DIFFERENTIATION**   1

Differentiation of a sum of terms — differentiation using 'function of a function' (or substitution) — differentiation of ax^n, sin ax, cos ax, $\log_e x$, e^{ax} — differentiation of products and quotients — differentiation of tan x — numerical values of differential coefficients.

Chapter 2   **SECOND DERIVATIVE**   24

First and second differential coefficients of expressions containing ax^n, sin ax, cos ax, tan x, $\log_e x$ and e^{ax} — velocity and acceleration as first and second derivatives of distance with respect to time.

Chapter 3   **MAXIMUM AND MINIMUM**   33

Turning points — tests for maximum and minimum — applications to graphs and problems in engineering.

Chapter 4   **AREAS AND VOLUMES BY INTEGRATION**   44

Integrals of the more common functions — the process of integration — application to finding areas — volumes of revolution.

Chapter 5   **THEOREMS OF PAPPUS FOR AREAS, VOLUMES AND CENTROIDS**   58
Theorem 1 concerning volume, area and centroid — Theorem 2 concerning area, length of arc and centroid.

Chapter 6   **CENTROIDS OF AREAS**   70

The first moment of area — formula for finding centroids — centroids by integration.

Chapter 7   **SECOND MOMENTS OF AREA**   84

The second moment of area from formula ΣAx^2 — second moment of area of a rectangle about its base — I_{XX} and I_{YY} of sectional areas which comprise rectangular areas — parallel axis theorem — polar second moment of area — perpendicular axis theorem — radius of gyration.

Chapter 8   **DIFFERENTIAL EQUATIONS**   109

Families of curves which represent equations of the type where $\dfrac{dy}{dx}$ is a function of x — meaning of a differential equation, a general solution, a boundary condition, and a particular solution — equations of the type $\dfrac{dy}{dx} = ky$ and their solutions.

Chapter 9   **THREE DIMENSIONAL TRIANGULATION**   119
            PROBLEMS

The solution of plane triangles — the angle between a line and a plane — the angle between two planes.

Chapter 10 **TRIGONOMETRICAL GRAPHS**   132

Amplitude or peak value — angular and time scales — cycle — period — frequency — phase angles of lead and lag.

Chapter 11 **COMPOUND ANGLE FORMULAE**   145

Formulae for $\sin(A+B)$, $\sin(A-B)$, $\cos(A+B)$, and $\cos(A-B)$ — the form $R.\sin(\theta \pm \alpha)$.

Chapter 12 **THE BINOMIAL THEOREM**   152

Binomial expansions of $(a+b)^n$ for small positive integer values of index n — Pascal's triangle for finding coefficients — the general form of the binomial theorem — application to small errors — approximations.

Chapter 13 **THE EXPONENTIAL FUNCTION**   158

The exponential function e^x expressed as a series — the numerical value of e — meaning of logarithms — logarithmic bases — natural logarithmic tables — exponential graphs of e^x and e^{-x} — growth and decay curves.

Chapter 14 **LOGARITHMIC GRAPH PAPER** 171

Non-linear laws which can be reduced to straight line form — the straight line form $y = mx + c$ — logarithmic scales — reduction of $z = a.t^n$, $z = a.b^t$, and $z = a.e^{bt}$ to straight line form — full logarithmic and semi-logarithmic graph paper.

Chapter 15 **PROBABILITY** 184

Simple probability — the probability scale — empirical probability — total probability — addition and multiplication laws — repeated trials — binomial coefficients — binomial distribution — the Poisson distribution.

Chapter 16 **THE NORMAL DISTRIBUTION** 210

Arithmetic mean and standard deviation — probability using areas under the normal curve — the normal distribution as an approximation to the binomial distribution for repeated trials — normal probability graph paper.

Chapter 17 **MOMENTS OF INERTIA** 228

Derivation of $I = Mr^2$ — radius of gyration k — moment of inertia of a solid cylinder — parallel axis theorem — moments of inertia of a solid cylinder and rectangular block about various axes.

Chapter 18 **COMPLEX NUMBERS** 236

Definition of j — algebraic form of a complex number — powers of j — addition, subtraction, multiplication and division of complex numbers in algebraic form — the Argand diagram — the j operator — addition and subtraction of phasors — polar form of a complex number — multiplying and dividing numbers in polar form.

ANSWERS 257

INDEX 261

AUTHORS' NOTE ON THE SERIES

Arising from the recommendations of the Haslegrave Report, the Technician Education Council has been set up. The Council has devised 'standard units', leading in various subjects to the award of its Certificates and Diplomas. The units are constructed at three levels, I, II and III.

A major change of emphasis in the educational approach adopted in T.E.C. Standard Units has been introduced by the use of 'objectives' throughout the courses, the intention being to allow student and lecturer to achieve planned progress through each unit on a step-by-step basis.

This set of books provides all the mathematics required for the T.E.C. Standard Units at each of the three levels. Each book follows a standard pattern, and each chapter opens with the words "After reaching the end of this chapter you should be able to:-" and this statement is followed by the objectives for that particular topic as laid down in the Standard Unit. Thereafter each chapter contains explanatory text, worked examples, and copious supplies of further exercises. As planned at present the series comprises:-

AN INTRODUCTORY COURSE	Level I (full unit)
MECHANICAL ENGINEERING MATHEMATICS	Level II (half-unit)
PRACTICAL MATHEMATICS	Level II (half-unit)
ANALYTICAL MATHEMATICS	Level II (half-unit)
ELECTRICAL ENGINEERING MATHEMATICS	Level II (full-unit)
ENGINEERING MATHEMATICS	Level III (full unit)

A. Greer
G. W. Taylor

Gloucester, 1978

NOTE ON THIS VOLUME

The authors wish to point out that although these topics are not listed specifically in the objectives of the Standard Unit, they have added sections in this book on Complex Numbers, and Moments of Inertia, believing that a large number of lecturers will welcome these additions. They also wish to record their appreciation of the valuable help and advice given in the preparation of the script by Mr. C. G. Crompton and others.

1. DIFFERENTIATION

On reaching the end of this chapter you should be able to :-

1. Use the derivatives of the functions : ax^n, sin ax, cos ax, tan x, log_e x, and e^{ax}.
2. Calculate the derivative at a point of the functions in 1.
3. State the basic rules of differential calculus for the derivatives of sum, product, quotient, and function of a function.
4. Determine the derivatives of various combinations of any two of the functions in 1 using 3.
5. Evaluate the derivatives in 4 at a given point.

DIFFERENTIATION

You may recall that the process of differentiation is a method of finding the rate of change of a function. The rate of change at a point on a curve may be found by determining the gradient of the tangent at that point. This may be achieved by either a theoretical or graphical method as follows.

Suppose we wish to find the gradient of the curve $y = x^2$ at the points where $x = 3$ and $x = -2$.

From a theoretical approach you may remember that if the function is of the form $y = x^n$, then the process of differentiating gives $\frac{dy}{dx} = n.x^{n-1}$.

Hence if
$$y = x^2$$

then
$$\frac{dy}{dx} = 2x$$

and therefore when $x = 3$, then the gradient of the tangent

$$\frac{dy}{dx} = 2(3) = +6$$

and when $x = -2$, then the gradient of the tangent

$$\frac{dy}{dx} = 2(-2) = -4.$$

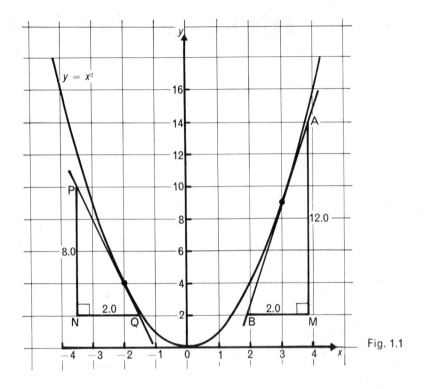

Fig. 1.1

Alternatively if we decide to use the graphical method we must first draw the graph as shown in Fig. 1.1.

Tangents have been drawn at the given points, that is where $x = 3$ and $x = -2$, and then the gradients may be found by constructing a suitable right angled triangle.

Where $\quad x = 3 \quad$ the gradient $= \dfrac{AM}{BM} = \dfrac{12.0}{2.0} = +6$

and where $\quad x = -2 \quad$ the gradient $= \dfrac{PN}{QN} = -\dfrac{8.0}{2.0} = -4.$

These results verify those obtained by the theoretical method.

DIFFERENTIATION OF A SUM

To differentiate an expression containing the sum of several terms, we differentiate each individual term separately.

Hence if $\quad y = 3x^2 + 3$

then $\quad \dfrac{dy}{dx} = 6x + 2$

And if $\qquad y = ax^3 + bx^2 - cx + d$

then $\qquad \dfrac{dy}{dx} = 3ax^2 + 2bx - c$

And if $\qquad y = \sqrt{x} + \dfrac{1}{\sqrt{x}} = x^{1/2} + x^{-1/2}$

then $\qquad \dfrac{dy}{dx} = \tfrac{1}{2}x^{-1/2} + (-\tfrac{1}{2})x^{-3/2} = \dfrac{1}{2\sqrt{x}} - \dfrac{1}{2\sqrt{x^3}}$

And if $\qquad y = 3.1x^{1.4} - \dfrac{3}{x} + 5 = 3.1x^{1.4} - 3x^{-1} + 5$

then $\qquad \dfrac{dy}{dx} = (3.1)(1.4)x^{0.4} - 3(-1)x^{-2} = 4.34x^{0.4} + \dfrac{3}{x^2}$

And if $\qquad y = \dfrac{t^3 + t}{t^2} = \dfrac{t^3}{t^2} + \dfrac{t}{t^2} = t + t^{-1}$

then $\qquad \dfrac{dy}{dt} = 1 + (-1)t^{-2} = 1 - \dfrac{1}{t^2}$

Exercise 1

1) Find $\dfrac{dy}{dx}$ if $y = 5x^3 + 7x^2 - x - 1$.

2) If $s = 7\sqrt{t} - 6t^{0.3}$ find an expression for $\dfrac{ds}{dt}$.

3) Find by a theoretical method the value of $\dfrac{dy}{dx}$ when $x = 2$ for the

curve $y = x - \dfrac{1}{x}$.

4) Find an expression for $\dfrac{dy}{du}$ if $y = \dfrac{u + u^2}{u}$.

5) Find graphically the gradient of the curve $y = x^2 + x + 2$ at the points where $x = +2$ and $x = -2$. Check the result by differentiation.

FUNCTIONS

If two variables x and y are connected so that the value of y depends upon the value allocated to x, then y is said to a *function* of x. Thus if $y = 3x^2 + 4x - 7$, when $x = 5$, $y = 3 \times 5^2 + 4 \times 5 - 7 = 88$. Since the value of y depends upon the value allocated to x, y is a function of x.

DIFFERENTIATION USING 'FUNCTION OF A FUNCTION'

Up to now we have only learnt how to differentiate comparatively simple expressions such as $y = 3x^{4.5}$. For more difficult expressions such as $y = \sqrt{(x^3+3x-9)}$ we make a substitution.

If we put $u = x^3+3x-9$ then $y = \sqrt{u}$.

Now y is a function of u, and since u is a function of x, it follows that y is a function of a function of x.

This all sounds rather complicated and the words 'function of a function' should be noted in case you meet them again, but we prefer to use the expression 'differentiation by substitution'.

DIFFERENTIATION BY SUBSTITUTION

A substitution method is often used for differentiating the more complicated expressions, together with the formula

$$\frac{dy}{dx} = \frac{dy}{du} \times \frac{du}{dx}$$

EXAMPLE 1

Find $\dfrac{dy}{dx}$ if $y = (x^2-x)^9$.

We have $\qquad\qquad y = (x^2-x)^9$

then $\qquad\qquad y = u^9 \qquad$ where $\quad u = x^2-x$

$\therefore \qquad\qquad \dfrac{dy}{du} = 9u^8 \qquad$ and $\quad \dfrac{du}{dx} = 2x-1$

but $\qquad\qquad \dfrac{dy}{dx} = \dfrac{dy}{du} \times \dfrac{du}{dx}$

$\therefore \qquad\qquad \dfrac{dy}{dx} = 9u^8 \times (2x-1)$

The differentiation has now been completed and it only remains to put u in terms of x by using our original substitution $u = x^2-x$

hence $\qquad\qquad \dfrac{dy}{dx} = 9(x^2-x)^8(2x-1)$

EXAMPLE 2

Find $\dfrac{d}{dx}\left(\sqrt{(1-5x^3)}\right)$

$\dfrac{d}{dx}\left(\sqrt{(1-5x^3)}\right)$ is called the differential coefficient of $\sqrt{(1-5x^3)}$ with respect to x. This simply means that we have to differentiate the expression with respect to x. If we let $y = \sqrt{(1-5x^3)}$ then the problem is to find $\dfrac{dy}{dx}$.

Let $\qquad\qquad y = \sqrt{(1-5x^3)}$

i.e. $\qquad\qquad y = (1-5x^3)^{1/2}$

then $\qquad\qquad y = u^{1/2} \qquad$ where $\quad u = 1-5x^3$

$\therefore\qquad\qquad \dfrac{dy}{du} = \tfrac{1}{2}u^{-1/2} \qquad$ and $\quad \dfrac{du}{dx} = -15x^2$

but $\qquad\qquad \dfrac{dy}{dx} = \dfrac{dy}{du}\times\dfrac{du}{dx}$

$\therefore\qquad\qquad \dfrac{dy}{dx} = \tfrac{1}{2}u^{-1/2}\times(-15x^2)$

$\therefore\qquad\qquad \dfrac{dy}{dx} = \tfrac{1}{2}(1-5x^3)^{-1/2}(-15x^2)$

$\therefore\qquad\qquad = -\dfrac{15}{2}x^2(1-5x^3)^{-1/2}$

$\qquad\qquad\qquad = -\dfrac{15x^2}{2\sqrt{(1-5x^3)}}$

Hence $\qquad \dfrac{d}{dx}\left(\sqrt{(1-5x^3)}\right) = -\dfrac{15x^2}{2\sqrt{(1-5x^3)}}$

DIFFERENTIATION OF A FUNCTION OF A FUNCTION BY RECOGNITION

Consider $y = (\quad)^n$ where any function of x can be written inside the bracket. Then differentiating with respect to x we have:

$$\frac{dy}{dx} = \frac{dy}{d(\quad)}\times\frac{d(\quad)}{dx}$$

Thus to differentiate an expression of the type $(\ )^n$, first differentiate the bracket, treating it as a term similar to x^n. Then differentiate the function of x inside the bracket. Finally, to obtain an expression for $\dfrac{dy}{dx}$, multiply these two results together.

EXAMPLE 3

Find $\dfrac{dy}{dx}$ if $y = (x^2 - 5x + 3)^5$

Differentiating the bracket as a whole we have:

$$\frac{dy}{d(\ )} = 5(x^2 - 5x + 3)^4 \qquad [1]$$

Also the function inside the bracket is $x^2 - 5x + 3$.

Differentiating this gives:

$$\frac{d(\ )}{dx} = 2x - 5 \qquad [2]$$

Thus multiplying the results [1] and [2] together gives:

$$\frac{dy}{dx} = 5(x^2 - 5x + 3)^4 \times (2x - 5)$$

$$= 5(2x - 5)(x^2 - 5x + 3)^4$$

Hence by recognising the method we can differentiate directly.

Consider, for example, if $\quad y = (x^2 - 3x)^7$

then $\quad \dfrac{dy}{dx} = 7(x^2 - 3x)^6 \times (2x - 3)$

$$= 7(2x - 3)(x^2 - 3x)^6$$

TO FIND THE RATE OF CHANGE OF sin θ, OR $\dfrac{d}{d\theta}\left(\sin\,\theta\right)$

The rate of change of a curve at any point is the gradient of the tangent at that point. We shall, therefore, find the gradient at various points on the graph of sin θ and then plot the values of these gradients to obtain a new graph.

It is suggested that the reader follows the method given, plotting his own curves on graph paper.

First, we plot the graph of $y = \sin\theta$ from $\theta = 0°$ to $\theta = 90°$ using values of $\sin\theta$ obtained from tables which are:

θ	$0°$	$15°$	$30°$	$45°$	$60°$	$75°$	$90°$
$y = \sin\theta$	0	0.259	0.500	0.707	0.866	0.966	1.000

These values are shown plotted in Fig. 1.2.

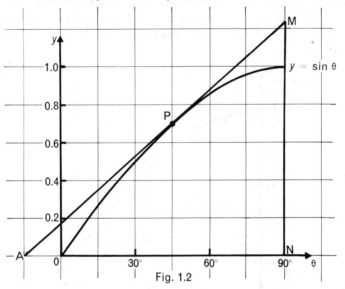

Fig. 1.2

Consider point P on the curve, where $\theta = 45°$, and draw the tangent APM.

We can find the gradient of the tangent by constructing a suitable right angled triangle AMN (which should be as large as conveniently possible for accuracy) and finding the value of $\dfrac{MN}{AN}$.

Using the scale on the y-axis MN = 1.29 by measurement, and using the scale on the θ-axis AN = 104° by measurement.

In calculations of this type it is necessary to obtain AN in radians. Remembering that

$$360° = 2\pi \text{ radians}$$

$$\therefore \qquad 1° = \frac{2\pi}{360} \text{ radians}$$

$$\therefore \qquad 104° = \frac{2\pi}{360} \times 104 = 1.81 \text{ radians}$$

Hence,

$$\text{the gradient at P} = \frac{MN}{AN} = \frac{1.29}{1.81} = 0.71$$

The value 0.71 is used as the y-value at $\theta = 45°$ to plot a point on a new graph using the same scales as before. This new graph could be plotted on the same axes as $y = \sin \theta$ but for clarity it has been shown on new axes in Fig. 1.3.

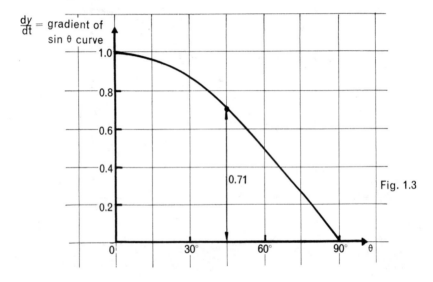

Fig. 1.3

This procedure is repeated for points on the $\sin \theta$ curve at θ values 0°, 15°, 30°, 60°, 75° and 90° and the new curve obtained will be as shown in Fig. 1.3. This is the graph of the gradients of the sine curve at various points.

If we now plot a graph of $\cos \theta$, taking values from tables, on the axes in Fig. 1.3 we shall find that the two curves coincide — any difference will be due to errors from drawing the tangents.

Hence the gradient of the $\sin \theta$ curve at any value of θ is the same as the value of $\cos \theta$.

In other words the rate of change of $\sin \theta$ is $\cos \theta$, provided that the angle θ is in radians.

In the above work we have only considered the graphs between 0° and 90° but the results are true for all values of the angle.

Hence if $\qquad y = \sin \theta$ then $\dfrac{dy}{dx} = \cos \theta$

or $\qquad \boxed{\dfrac{d}{d\theta}\Big(\sin \theta \Big) = \cos \theta}$ $\qquad$ provided that θ is in radians.

The same procedure may be used to show that:

$$\boxed{\dfrac{d}{d\theta}\Big(\cos \theta \Big) = -\sin \theta}$$ $\qquad$ provided that θ is in radians.

EXAMPLE 4

Find $\dfrac{d}{d\theta}\Big(\sin 7\theta \Big).$

Let $\qquad\qquad y = \sin 7\theta$

then $\qquad\qquad y = \sin u \qquad\qquad$ where $\quad u = 7\theta$

$\therefore \qquad\qquad \dfrac{dy}{du} = \cos u \qquad\qquad$ and $\quad \dfrac{du}{d\theta} = 7$

but $\qquad\qquad \dfrac{dy}{d\theta} = \dfrac{dy}{du} \times \dfrac{du}{d\theta}$

$\therefore \qquad\qquad \dfrac{dy}{d\theta} = (\cos u) \times 7 = 7 \cos u = 7 \cos 7\theta$

$\therefore \qquad \dfrac{d}{d\theta}\Big(\sin 7\theta \Big) = 7 \cos 7\theta$

EXAMPLE 5

Find $\dfrac{d}{d\theta}\Big(\cos 4\theta. \Big)$

Let $\qquad\qquad y = \cos 4\theta$

then $\qquad\qquad y = \cos u \qquad\qquad$ where $\quad u = 4\theta$

$\therefore \qquad\qquad \dfrac{dy}{du} = -\sin u \qquad\qquad$ and $\quad \dfrac{du}{d\theta} = 4$

but $\qquad\qquad \dfrac{dy}{d\theta} = \dfrac{dy}{du} \times \dfrac{du}{d\theta}$

$$\therefore \qquad \frac{dy}{d\theta} = (-\sin u) \times 4 = -4 \sin u = -4 \sin 4\theta$$

$$\therefore \qquad \frac{d}{d\theta}\left(\cos 4\theta\right) = -4 \sin 4\theta$$

In general

$$\boxed{\begin{array}{l} \dfrac{d}{dx}\left(\sin ax\right) = a \cos ax \\[2mm] \dfrac{d}{dx}\left(\cos ax\right) = -a \sin ax \end{array}}$$

and

EXAMPLE 6

Find $\dfrac{d}{dt}\left\{\cos\left(2t-\dfrac{3\pi}{2}\right)\right\}$

Let $\qquad\qquad y = \cos\left(2t-\dfrac{3\pi}{2}\right)$

then $\qquad\qquad y = \cos u \qquad\qquad$ where $\quad u = 2t-\dfrac{3\pi}{2}$

$$\therefore \qquad \frac{dy}{du} = -\sin u \qquad\qquad \text{and} \quad \frac{du}{dt} = 2$$

but $\qquad\qquad \dfrac{dy}{dt} = \dfrac{dy}{du}\times\dfrac{du}{dt}$

$$\therefore \qquad \frac{dy}{dt} = (-\sin u)\times 2 = -2 \sin u = -2 \sin\left(2t-\frac{3\pi}{2}\right)$$

$$\therefore \quad \frac{d}{dt}\left\{\cos\left(2t-\frac{3\pi}{2}\right)\right\} = -2 \sin\left(2t-\frac{3\pi}{2}\right)$$

THE DIFFERENTIAL COEFFICIENT OF $\log_e x$ OR $\frac{d}{dx}(\log_e x)$

In higher mathematics all logarithms are taken to the base e, where e = 2.71828. Logarithms to this base are often called natural logarithms. They are also called Naperian or hyperbolic logarithms and are given as $\log_e$ or ln.

Again the graphical method of differentiation may be used. Fig. 1.4 shows the graph of $\log_e x$.

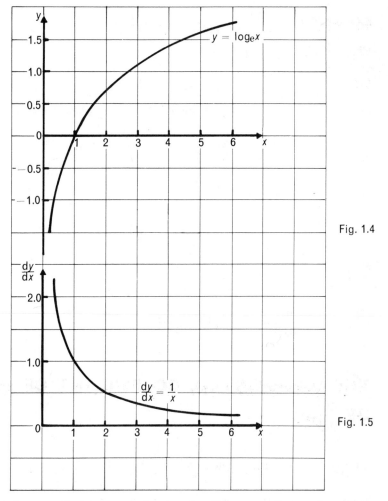

Fig. 1.4

Fig. 1.5

The reader may find it instructive to plot the curve of $y = \log_e x$ as shown in Fig. 1.4 and follow the procedure, as used previously, of drawing tangents at various points. The values of their gradients are then plotted and a curve will result as shown in Fig. 1.5.

If a graph of $\dfrac{1}{x}$ is plotted on the same axes as in Fig. 1.5 it will be found

to coincide with the gradient curve, that is, the $\dfrac{dy}{dx}$ graph.

Hence by differentiating $\log_e x$ we obtain $\dfrac{1}{x}$.

Hence

$$\boxed{\dfrac{d}{dx}\left(\log_e x\right) = \dfrac{1}{x}}$$

EXAMPLE 7

Find $\dfrac{d}{dx}\left(\log_e(x^2+5)\right)$.

Let $\qquad\qquad\qquad\qquad y = \log_e(x^2+5)$

then $\qquad\qquad\qquad\qquad y = \log_e u \qquad\qquad$ where $\quad u = x^2+5$

∴ $\qquad\qquad\qquad\qquad \dfrac{dy}{du} = \dfrac{1}{u} \qquad\qquad\qquad$ and $\quad \dfrac{du}{dx} = 2x$

but $\qquad\qquad\qquad\qquad \dfrac{dy}{dx} = \dfrac{dy}{du} \times \dfrac{du}{dx}$

∴ $\qquad\qquad\qquad\qquad \dfrac{dy}{dx} = \dfrac{1}{u} \times 2x = \dfrac{1}{x^2+5} \times 2x$

∴ $\qquad\qquad\qquad\qquad \dfrac{dy}{dx} = \dfrac{2x}{x^2+5}$

hence $\qquad\quad \dfrac{d}{dx}\left(\log_e(x^2+5)\right) = \dfrac{2x}{x^2+5}$

THE DIFFERENTIAL COEFFICIENT OF e^x, OR $\dfrac{d}{dx}\left(e^x\right)$

If we let $\quad y = e^x \quad$ then we need to find $\dfrac{dy}{dx}$.

Then $\qquad\qquad\qquad\qquad e^x = y$

and rearranging in log form we get $\quad x = \log_e y$

∴ $\qquad\qquad\qquad\qquad \dfrac{dx}{dy} = \dfrac{1}{y}$

hence by inverting both sides $\qquad \dfrac{dy}{dx} = y$

but we know that $\quad y = e^x \quad$ and ∴ $\dfrac{dy}{dx} = e^x$

Hence $\qquad\qquad\qquad\qquad \dfrac{d}{dx}\left(e^x\right) = e^x$

EXAMPLE 8

Find $\dfrac{d}{dx}\left(e^{6x}\right)$

Let $y = e^{6x}$

then $y = e^u$ where $u = 6x$

$\therefore$ $\dfrac{dy}{du} = e^u$ and $\dfrac{du}{dx} = 6$

but $\dfrac{dy}{dx} = \dfrac{dy}{du} \times \dfrac{du}{dx}$

$\therefore$ $\dfrac{dy}{dx} = e^u \times 6 = e^{6x} \times 6$

$\therefore$ $\dfrac{dy}{dx} = 6e^{6x}$

hence $\dfrac{d}{dx}\left(e^{6x}\right) = 6e^{6x}$

EXAMPLE 9

Find $\dfrac{d}{dx}\dfrac{1}{e^{3x}}$

Let $y = \dfrac{1}{e^{3x}}$

or $y = e^{-3x}$

then $y = e^u$ where $u = -3x$

$\therefore$ $\dfrac{dy}{du} = e^u$ and $\dfrac{du}{dx} = -3$

but $\dfrac{dy}{dx} = \dfrac{dy}{du} \times \dfrac{du}{dx}$

$\therefore$ $\dfrac{dy}{dx} = e^u \times (-3) = -3e^u = -3e^{-3x}$

$\therefore$ $\dfrac{d}{dx}\left(\dfrac{1}{e^{3x}}\right) = \dfrac{d}{dx}\left(e^{-3x}\right) = -3e^{-3x}$

In general

$$\boxed{\dfrac{d}{dx}\left(e^{ax}\right) = a.e^{ax}}$$

We may now summarise differential coefficients of the more common functions:

y	$\dfrac{dy}{dx}$
$a.x^n$	$an.x^{n-1}$
$\sin ax$	$a.\cos ax$
$\cos ax$	$-a.\sin ax$
$\log_e x$	$\dfrac{1}{x}$
e^{ax}	$a.e^{ax}$

At this stage we may use the results in the table to differentiate many functions by 'recognition' rather than using the method of substitution.

For example
$$\frac{d}{d\theta}(\sin 3\theta) = 3.\cos 3\theta$$

or
$$\frac{d}{dx}(e^{-3x}) = -3e^{-3x}$$

However, if the functions to be differentiated are not similar to those shown in the table, perhaps because they are more complicated, the method of substitution should be used.

Example 10 which follows shows how both the method of substitution and recognition may be used together to differentiate a more complicated function.

EXAMPLE 10

Find $\dfrac{dy}{dx}$ if $y = \sin^3 5x$.

We have $\qquad y = (\sin 5x)^3$

then $\qquad y = u^3 \qquad\qquad$ where $\quad u = \sin 5x$

$\therefore \qquad \dfrac{dy}{du} = 3u^2 \qquad\qquad$ and $\quad \dfrac{du}{dx} = 5\cos 5x$

but $$\frac{dy}{dx} = \frac{dy}{du} \times \frac{du}{dx}$$

$\therefore$ $$\frac{dy}{dx} = 3u^2 \times 5 \cos 5x = 3 \sin^2 5x \times 5 \cos 5x$$

$\therefore$ $$\frac{dy}{dx} = 15 \sin^2 5x \cos 5x$$

Exercise 2

Differentiate with respect to x:

1) $(3x+1)^2$

2) $(2-5x)^3$

3) $(1-4x)^{1/2}$

4) $(2-5x)^{3/2}$

5) $\dfrac{1}{4x^2+3}$

6) $\sin(3x+4)$

7) $\cos(2-5x)$

8) $\sin^2 4x$

9) $\dfrac{1}{\cos^3 7x}$

10) $\sin\left(2x+\dfrac{\pi}{2}\right)$

11) $\cos^3 x$

12) $\dfrac{1}{\sin x}$

13) $\log_e 9x$

14) $9.\log_e\left(\dfrac{5}{x}\right)$

15) $\dfrac{1}{4}\log_e(2x-7)$

16) $\dfrac{1}{e^x}$

17) $2.e^{3x+4}$

18) $\dfrac{1}{e^{2-8x}}$

19) Find $\dfrac{d}{dt}\left(\dfrac{1}{\sqrt[3]{1-2t}}\right)$

20) Find $\dfrac{d}{d\theta}\left(\sin\left(\tfrac{3}{4}\theta-\pi\right)\right)$

21) Find $\dfrac{d}{d\phi}\left(\dfrac{1}{\cos(\pi-\phi)}\right)$

22) Find $\dfrac{d}{dx}\left(\log_e\dfrac{1}{\sqrt{x}}\right)$

23) Find $\dfrac{d}{dt}\left(B.e^{kt-b}\right)$

24) Find $\dfrac{d}{dx}\left(\sqrt[3]{e^{1-x}}\right)$

DIFFERENTIATION OF A PRODUCT

If $y = u \times v$ where u and v are functions of x, we must use the formula:

$$\frac{dy}{dx} = v.\frac{du}{dx} + u.\frac{dv}{dx}$$

EXAMPLE 11

Find $\dfrac{d}{dx}\left(x^3\sin 2x\right)$.

Let $\qquad\qquad y = x^3\sin 2x$

then $\qquad\qquad y = u\times v \quad$ where $\quad u = x^3 \quad$ and $\quad v = \sin 2x$

$$\therefore \qquad \frac{du}{dx} = 3x^2 \quad \text{and} \quad \frac{dv}{dx} = 2\cos 2x$$

but $\qquad\qquad \dfrac{dy}{dx} = v.\dfrac{du}{dx}+u.\dfrac{dv}{dx}$

$\therefore \qquad\qquad \dfrac{dy}{dx} = (\sin 2x)3x^2+x^3(2\cos 2x)$

$$= 3x^2\sin 2x+2x^3\cos 2x$$

$$= x^2(3\sin 2x+2x\cos 2x)$$

then $\qquad \dfrac{d}{dx}\left(x^3\sin 2x\right) = x^2(3\sin 2x+2x\cos 2x)$

EXAMPLE 12

Differentiate $(x^2+1)\log_e x$ with respect to x.

If we let $y = (x^2+1)\log_e x$ then the problem is to find $\dfrac{dy}{dx}$.

Then $\qquad y = u\times v \quad$ where $\quad u = x^2+1 \quad$ and $\quad v = \log_e x$

$$\therefore \qquad \frac{du}{dx} = 2x \qquad \text{and} \qquad \frac{dv}{dx} = \frac{1}{x}$$

but $\qquad\qquad \dfrac{dy}{dx} = v.\dfrac{du}{dx}+u.\dfrac{dv}{dx}$

$\therefore \qquad\qquad \dfrac{dy}{dx} = (\log_e x)2x+(x^2+1)\dfrac{1}{x}$

$$= 2x(\log_e x)+x+\frac{1}{x}$$

DIFFERENTIATION OF A QUOTIENT

If $y = \dfrac{u}{v}$ where u and v are functions of x, we must use the formula

$$\boxed{\dfrac{dy}{dx} = \dfrac{v \cdot \dfrac{du}{dx} - u \cdot \dfrac{dv}{dx}}{v^2}}$$

EXAMPLE 13

Find $\dfrac{dy}{dx}$ if $y = \dfrac{e^{2x}}{x+3}$

We have $\qquad y = \dfrac{e^{2x}}{x+3}$

Let $\qquad y = \dfrac{u}{v}$ where $u = e^{2x}$ and $v = x+3$

$$\therefore \quad \dfrac{du}{dx} = 2e^{2x} \quad \text{and} \quad \dfrac{dv}{dx} = 1$$

but $\qquad \dfrac{dy}{dx} = \dfrac{v \cdot \dfrac{du}{dx} - u \cdot \dfrac{dv}{dx}}{v^2}$

$$\therefore \quad \dfrac{dy}{dx} = \dfrac{(x+3)2e^{2x} - e^{2x} \times 1}{(x+3)^2}$$

$$= \dfrac{e^{2x}(2x+6-1)}{(x+3)^2}$$

$$= \dfrac{(2x+5)e^{2x}}{(x+3)^2}$$

EXAMPLE 14

Find $\dfrac{d}{d\theta}\left(\tan\theta\right)$

Let $\qquad y = \tan\theta$

$$\therefore \qquad y = \dfrac{\sin\theta}{\cos\theta}$$

then $\qquad y = \dfrac{u}{v}$ where $u = \sin\theta$ and $v = \cos\theta$

$$\therefore \quad \dfrac{du}{d\theta} = \cos\theta \quad \text{and} \quad \dfrac{dv}{d\theta} = -\sin\theta$$

but
$$\frac{dy}{d\theta} = \frac{v.\dfrac{du}{d\theta} - u.\dfrac{dv}{d\theta}}{v^2}$$

$\therefore$
$$\frac{dy}{d\theta} = \frac{(\cos\theta)(\cos\theta) - (\sin\theta)(-\sin\theta)}{\cos^2\theta}$$

$$= \frac{\cos^2\theta + \sin^2\theta}{\cos^2\theta}$$

Using the identity $\sin^2\theta + \cos^2\theta = 1$.

then
$$\frac{dy}{d\theta} = \frac{1}{\cos^2\theta} = \sec^2\theta$$

Hence $\boxed{\dfrac{d}{d\theta}\left(\tan\theta\right) = \sec^2\theta}$ provided that θ is in radians.

We may now summarise differential coefficients of the more common functions:

y	$\dfrac{dy}{dx}$
$a.x^n$	$an.x^{n-1}$
$\sin ax$	$a.\cos ax$
$\cos ax$	$-a.\sin ax$
$\tan x$	$\sec^2 x$
$\log_e x$	$\dfrac{1}{x}$
e^{ax}	$a.e^{ax}$

Exercise 3

1) Differentiate with respect to x:

(a) $x.\sin x$ (b) $e^x.\tan x$ (c) $x.\log_e x$

2) Find $\dfrac{d}{dt}\left(\sin t.\cos t\right)$ **3)** Find $\dfrac{d}{d\theta}\left(\sin 2\theta.\tan\theta\right)$

4) Find $\dfrac{d}{dm}\left(e^{4m}\cos 3m\right)$

6) Find $\dfrac{d}{dt}\left(6e^{3t}(t^2-1)\right)$

5) Find $\dfrac{d}{dx}\left(3x^2\log_e x\right)$

7) Find $\dfrac{d}{dz}\left((z-3z^2)\log_e z\right)$

8) Differentiate with respect to x:

(a) $\dfrac{x}{1-x}$

(b) $\dfrac{\log_e x}{x^2}$

(c) $\dfrac{e^x}{\sin 2x}$

9) Find $\dfrac{d}{dz}\left(\dfrac{z+2}{3-4z}\right)$

10) Find $\dfrac{d}{dt}\left(\dfrac{\cos 2t}{e^{2t}}\right)$

11) Find $\dfrac{d}{d\theta}\left(\cot\theta\right)$ $\left(\textit{Hint:}\ \ \text{use the identity}\ \ \cot\theta=\dfrac{\cos\theta}{\sin\theta}\right)$

NUMERICAL VALUES OF DIFFERENTIAL COEFFICIENTS

EXAMPLE 15

Find the value of $\dfrac{dy}{dx}$ for the curve $y=\dfrac{1}{\sqrt{x}}-3.\log_e x$ at the point where $x=2.3$.

We have
$$y=x^{-1/2}-3.\log_e x$$

$\therefore$
$$\frac{dy}{dx}=-\tfrac{1}{2}x^{-3/2}-\frac{3}{x}$$

It is often difficult to decide how much simplification of an expression will help in finding its numerical value when a particular value of x is substituted. In this case the expression for $\dfrac{dy}{dx}$ may be rewritten as

$$\frac{dy}{dx}=-\frac{1}{2(\sqrt{x})^3}-\frac{3}{x}.$$

Hence when $x=2.3$

then
$$\frac{dy}{dx}=-\frac{1}{2(\sqrt{2.3})^3}-\frac{3}{2.3}=-1.45$$

It may well be argued that if a scientific calculator is available the value of x may just as well be substituted into the original expression for $\dfrac{dy}{dx}$ without any simplification.

This would give $\dfrac{dy}{dx} = -\tfrac{1}{2}(2.3)^{-1.5} - \dfrac{3}{2.3}$ which can be evaluated as easily as the 'simplified' arrangement.

It may, however, be more difficult to detect a computation error when making a rough check of the answer which is why expressions with positive indices are often preferred.

EXAMPLE 16

A curve is given in the form $y = 3 \sin 2\theta - 5 \tan \theta$ when θ is in radians. Find the gradient of the curve at the point where θ has a value equivalent to 34°.

The gradient of the curve is given by $\dfrac{dy}{d\theta}$.

We have $y = 3 \sin 2\theta - 5 \tan \theta$

∴ $\dfrac{dy}{d\theta} = 3 \times 2 \cos 2\theta - 5 \sec^2\theta$

and substituting the value $\theta = 34°$

then $\dfrac{dy}{d\theta} = 6 \cos (2 \times 34)° - 5 \sec^2(34)°$

$= 6 \cos 68° - 5(\sec 34°)^2$

$= -5.03$

EXAMPLE 17

If $y = \dfrac{1}{2}(e^{3t} + e^{-3t})$ find the value of $\dfrac{dy}{dt}$ when $t = -0.63$.

We have $y = \dfrac{1}{2}\left[e^{3t} + e^{-3t}\right]$

∴ $\dfrac{dy}{dt} = \dfrac{1}{2}\left[3e^{3t} + (-3)e^{-3t}\right]$

And substituting the value $t = -0.63$

then $\dfrac{dy}{dt} = \dfrac{3}{2}\left[e^{3(-0.63)} - e^{-3(-0.63)}\right]$

$= \dfrac{3}{2}\left[e^{-1.89} - e^{1.89}\right]$

$= \dfrac{3}{2}\left[0.151 - 6.619\right]$

$= -9.70$

EXAMPLE 18

Find the gradient of the curve $\dfrac{\cos x}{x}$ at the point where $x = 0.25$.

The gradient of the curve is given by $\dfrac{dy}{dx}$ if we let,

$$y = \frac{\cos x}{x}.$$

Then $\qquad y = \dfrac{u}{v}$ where $u = \cos x$ and $v = x$

$$\therefore \qquad \frac{du}{dx} = -\sin x \quad \therefore \quad \frac{dv}{dx} = 1$$

but $\qquad \dfrac{dy}{dx} = \dfrac{v.\dfrac{du}{dx} - u.\dfrac{dy}{dx}}{v^2}$

$\therefore \qquad \dfrac{dy}{dx} = \dfrac{x(-\sin x) - (\cos x)1}{x^2}$

$$= \frac{-x.\sin x - \cos x}{x^2}$$

If we now substitute the value $x = 0.25$

then $\qquad \dfrac{dy}{dx} = \dfrac{-(0.25)(\sin 0.25) - (\cos 0.25)}{(0.25)^2}$

The value 0.25 must be treated as radians when substituted into trigonometrical functions such as $\sin x$ and $\cos x$.

If a scientific calculating machine is used it is usually possible to set the machine to accept radians and give directly trigonometrical ratios.

then $\qquad \dfrac{dy}{dx} = -16.5$

If tables are to be used to perform the calculation it may well be necessary to convert the angle to degrees.

Now $\quad 0.25 \text{ rad} = \left(0.25 \times \dfrac{180}{\pi}\right)^{\circ} = 14° \ 19'$

and hence $\qquad \dfrac{dy}{dx} = \dfrac{-(0.25)\sin 14° \ 19' - \cos 14° \ 19'}{(0.25)^2}$

$$= -16.5$$

Exercise 4

1) If $y = 3x^2 - \dfrac{7}{x^2} + \sqrt{x}$ find the value of $\dfrac{dy}{dx}$ if $x = 3.5$.

2) If $y = 5 \sin 2\theta + 3 \cos \dfrac{\theta}{2}$ find the value of $\dfrac{dy}{d\theta}$ if $\theta = 0.942$ radians.

3) Find the value of $\dfrac{dy}{dt}$ when $t = -0.1$ if $y = \dfrac{1}{2}(e^t - e^{-t})$.

4) If $x = 0.3$ find the value of $\dfrac{dy}{dx}$ when $y = \sqrt{(3 - 2x^2)}$.

5) If $y = \sin^4 2\theta$ find the value of $\dfrac{dy}{d\theta}$ when $\theta = \dfrac{3\pi}{2}$ radians.

6) A curve is given in the form $y = \tan(3\phi - \pi)$, where ϕ is in radians. Find the gradient of the curve at the point where ϕ has a value equivalent to $23.4°$.

7) If $y = 4 \log_e(1 - x)$ find the value of $\dfrac{dy}{dx}$ when $x = 0.32$.

8) Find the value of $\dfrac{dy}{dx}$ when $x = 2.9$ when $y = e^{(9 - 3x)}$.

9) Given that $y = (\sin x)(\cos x)$ find the value of $\dfrac{dy}{dx}$ when $x = \dfrac{\pi}{6}$ radians.

10) If $y = \dfrac{1 + x^2}{x - 2}$ find the value of $\dfrac{dy}{dx}$ if $x = -1.25$.

SUMMARY

a) The table shows differential coefficients of the more common functions:

y	$\dfrac{dy}{dx}$
$a.x^n$	$an.x^{n-1}$
$\sin ax$	$a.\cos ax$
$\cos ax$	$-a.\sin ax$
$\tan x$	$\sec^2 x$
$\log_e x$	$\dfrac{1}{x}$
e^{ax}	$a.e^{ax}$

b) Differentiation by substitution involves use of the formula

$$\frac{dy}{dx} = \frac{dy}{du} \times \frac{du}{dx}.$$

c) Differentiation of a product:

If $y = u \times v$ then $\dfrac{dy}{dx} = v.\dfrac{du}{dx} + u.\dfrac{dv}{dx}.$

d) Differentiation of a quotient:

If $y = \dfrac{u}{v}$ then $\dfrac{dy}{dx} = \dfrac{v.\dfrac{du}{dx} - u.\dfrac{dv}{dx}}{v^2}$

SECOND DERIVATIVE

On reaching the end of this chapter you should be able to:-

1. State the notation for second derivatives as $\dfrac{d^2y}{dx^2}$ and similar form e.g. $\dfrac{d^2x}{dt^2}$

2. Determine a second derivative, by applying the basic rules of differential calculus, to the simplified result of a first differentiation.

3. Evaluate a second derivative determined in 2 at a given point.

4. State that $\dfrac{ds}{dt}$ and $\dfrac{d^2s}{dt^2}$ express velocity and acceleration.

5. Calculate the velocity and acceleration at a given time from an equation for displacement expressed in terms of time using 4.

SECOND DERIVATIVE

If
$$y = x^6$$

then
$$\frac{dy}{dx} = 6x^5$$

and if we differentiate this equation again with respect to x we obtain:

$$\frac{d}{dx}\left(\frac{dy}{dx}\right) = \frac{d}{dx}(6x^5)$$

or
$$\frac{d^2y}{dx^2} = 30x^4$$

Now just as $\dfrac{dy}{dx}$ is called the first differential coefficient, or first derivative, of y with respect to x, then $\dfrac{d^2y}{dx^2}$ is called the second differential coefficient, or second derivative, of y with respect to x.

It should be noted that the figure 2 which occurs twice in $\dfrac{d^2y}{dx^2}$ is NOT an index but merely indicates that the original function has been differentiated twice. Hence $\dfrac{d^2y}{dx^2}$ is NOT the same as $\left(\dfrac{dy}{dx}\right)^2$.

EXAMPLE 1

If $y = x^3 - 2x^2 + 3x - 7$ find $\dfrac{dy}{dx}$ and $\dfrac{d^2y}{dx^2}$.

Now $$y = x^3 - 2x^2 + 3x - 7$$

$\therefore$ $$\frac{dy}{dx} = 3x^2 - 4x + 3$$

and $$\frac{d^2y}{dx^2} = 6x - 4$$

EXAMPLE 2

If $y = \sqrt{x} + \log_e x$ find the values of $\dfrac{dy}{dx}$ and $\dfrac{d^2y}{dx^2}$ when $x = 2$.

Now $$y = x^{1/2} + \log_e x$$

$\therefore$ $$\frac{dy}{dx} = \tfrac{1}{2}x^{-1/2} + \frac{1}{x}$$

$$= \tfrac{1}{2}x^{-1/2} + x^{-1}$$

$\therefore$ $$\frac{d^2y}{dx^2} = -\tfrac{1}{4}x^{-3/2} + (-1)x^{-2} = -\tfrac{1}{4}x^{-3/2} - x^{-2}$$

Before substituting the numerical value of x we recommend that the expressions for $\dfrac{dy}{dx}$ and $\dfrac{d^2y}{dx^2}$ are transformed to give positive indices.

Although most electronic calculating machines will evaluate expressions with negative indices mistakes may occur and it is more difficult to spot a computation error.

Now $$\frac{dy}{dx} = \frac{1}{2x^{1/2}} + \frac{1}{x} = \frac{1}{2\sqrt{x}} + \frac{1}{x}$$

and hence when $x = 2$ then,

$$\frac{dy}{dx} = \frac{1}{2\sqrt{2}} + \frac{1}{2} = 0.354 + 0.5 = 0.854$$

also $$\frac{d^2y}{dx^2} = -\frac{1}{4x^{3/2}} - \frac{1}{x^2} = -\frac{1}{4(\sqrt{x})^3} - \frac{1}{x^2}$$

and when $x = 2$ then,

$$\frac{d^2y}{dx^2} = -\frac{1}{4(\sqrt{2})^3} - \frac{1}{2^2} = -0.088 - 0.250 = -0.338$$

EXAMPLE 3

If $\theta = \dfrac{\pi}{2}$ find the values of $\dfrac{dy}{d\theta}$ and $\dfrac{d^2y}{d\theta^2}$ given that $y = \sin 2\theta + \cos 3\theta$.

Now $y = \sin 2\theta + \cos 3\theta$

$\therefore$ $\dfrac{dy}{d\theta} = 2\cos 2\theta - 3\sin 3\theta$

and $\dfrac{d^2y}{d\theta^2} = -4\sin 2\theta - 9\cos 3\theta$

If a value of an angle is given in terms of π the units are radians.

$\therefore$ when $\theta = \dfrac{\pi}{2}$ we have:

$$\dfrac{dy}{d\theta} = 2\cos 2\left(\dfrac{\pi}{2}\right) - 3\sin 3\left(\dfrac{\pi}{2}\right)$$

$$= 2\cos 180° - 3\sin 270° = 2(-1) - 3(-1) = 1$$

and when $\theta = \dfrac{\pi}{2}$ we have:

$$\dfrac{d^2y}{d\theta^2} = -4\sin 2\left(\dfrac{\pi}{2}\right) - 9\cos 3\left(\dfrac{\pi}{2}\right) = -4(0) - 9(0) = 0$$

EXAMPLE 4

If $y = \frac{1}{2}(e^{2t} + e^{-2t})$ find the values of $\dfrac{dy}{dt}$ and $\dfrac{d^2y}{dt^2}$ if $t = 0.61$.

We have, $y = \frac{1}{2}e^{2t} + \frac{1}{2}e^{-2t}$

$\therefore$ $\dfrac{dy}{dt} = \frac{1}{2}(2e^{2t}) + \frac{1}{2}(-2e^{-2t})$

$$= e^{2t} - e^{-2t}$$

Also $\dfrac{d^2y}{dt^2} = 2e^{2t} - (-2)e^{-2t}$

$$= 2e^{2t} + 2e^{-2t}$$

hence when $t = 0.61$ then

$$\dfrac{dy}{dt} = e^{2(0.61)} - e^{-2(0.61)} = 3.09$$

and when $t = 0.61$ then

$$\dfrac{d^2y}{dt^2} = 2e^{2(0.61)} + 2e^{-2(0.61)} = 7.36$$

EXAMPLE 5

If $y = \tan\theta$ find the value of $\dfrac{d^2y}{d\theta^2}$ when $\theta = 0.436$ radians.

We have, $$y = \tan\theta$$

$$\therefore \qquad \frac{dy}{d\theta} = \sec^2\theta$$

that is $$\frac{dy}{d\theta} = (\cos\theta)^{-2}$$

$$\therefore \qquad \frac{d^2y}{d\theta^2} = 2(\sin\theta)(\cos\theta)^{-3}$$

The reader is left to check this second differentiation by differentiating $(\cos\theta)^{-2}$ by the method of substitution.

Now when $\theta = 0.436$ radians then

$$\frac{d^2y}{d\theta^2} = \frac{2(\sin 0.436)}{(\cos 0.436)^3} = 1.135$$

Exercise 5

1) If $y = 3x^3 + 2x - 7$ find an expression for $\dfrac{d^2y}{dx^2}$ and also its value when $x = 3$.

2) Find the value of $\dfrac{dy}{dx}$ and $\dfrac{d^2y}{dx^2}$ when $x = -2$ given that $y = 5x^4 + 7x^2 + x$.

3) Given that $y = \dfrac{3t^5 + 2t}{t^2}$ find the value of $\dfrac{d^2y}{dt^2}$ when $t = 0.6$.

4) If $y = \log_e x$ find $\dfrac{d^2y}{dx^2}$ in terms of x, and also its value when $x = 1.9$.

5) If $z = 5\cos 3\theta$ find the value of $\dfrac{dz}{d\theta}$ and $\dfrac{d^2z}{d\theta^2}$ when $\theta = \dfrac{\pi}{2}$ radians.

6) Find the value of $\dfrac{d^2p}{d\phi^2}$ given that $p = 6\sin 4\phi$, where ϕ is in radians, when ϕ has a value equivalent to $28°$.

7) If $y = 2\cos\dfrac{\alpha}{4}$ find the value of $\dfrac{d^2y}{d\alpha^2}$ when $\alpha = 1.6$ radians.

8) When $t = 2$ find the value of $\dfrac{d^2v}{dt^2}$ given that $v = e^t + e^{-t}$.

9) If $y = -e^{4.6t}$ find the value of $\dfrac{dy}{dt}$ and $\dfrac{d^2y}{dt^2}$ when $t = 0$.

10) Find the value of $\dfrac{d^2u}{dm^2}$ if $u = \dfrac{1}{2}(e^{3m} - e^{-3m})$ given that $m = 1.3$.

11) If $y = x.\log_e x$ find the value of $\dfrac{d^2y}{dx^2}$ if $x = 0.34$.

VELOCITY AND ACCELERATION

Suppose that a vehicle starts from rest and travels 60 metres in 12 seconds. The average velocity may be found by dividing the total distance travelled by the total time taken, that is $\dfrac{60}{12} = 5$ m/s. This is NOT the INSTANTANEOUS velocity, however, AT a time of 12 seconds, but is the AVERAGE VELOCITY over the 12 seconds as calculated previously.

Fig. 2.1 shows a graph of distance s, against time t. The average velocity over a period is given by the gradient of the chord which meets the curve at the extremes of the period. Thus in the diagram the gradient of the dotted chord QR gives the average velocity between $t = 2$ s and $t = 6$ s. It is found to be $\frac{13}{4} = 3.25$ m/s.

The velocity at any point is the rate of change of s with respect t and may be found by finding the gradient of the curve at that point. In mathematical notation this is given by $\dfrac{ds}{dt}$.

Suppose we know that the relationship between s and t is

$$s = 0.417t^2$$

then velocity, $v = \dfrac{ds}{dt} = 0.834t$

and hence when $t = 12$ seconds, then $v = 0.834 \times 12 = 10$ m/s.

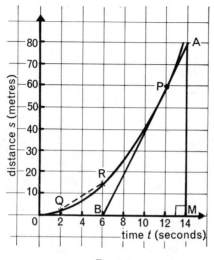

Fig. 2.1

This result may be found graphically by drawing the tangent to the curve of s against t at the point P and constructing a suitable right angled triangle ABM.

Hence the velocity at $\quad P = \dfrac{AM}{BM} = \dfrac{80}{8} = 10$ m/s   which   verifies   the theoretical result.

Similarly, the rate of change of velocity with respect to time is called acceleration and is given by the gradient of the velocity–time graph at any point. In mathematical notation this is given by $\dfrac{dv}{dt}$.

Now

$$\frac{dv}{dt} = \frac{d}{dt}(v) = \frac{d}{dt}\left(\frac{ds}{dt}\right) = \frac{d^2s}{dt^2}$$

and so the acceleration, a, is given by either

$$\frac{dv}{dt} \quad \text{or} \quad \frac{d^2s}{dt^2}.$$

The above reasoning was applied to linear motion, but it could also have been used for angular motion. The essential difference is that distance, s, is replaced by angle turned through, θ rad.

Both sets of results are summarised in Fig. 2.2.

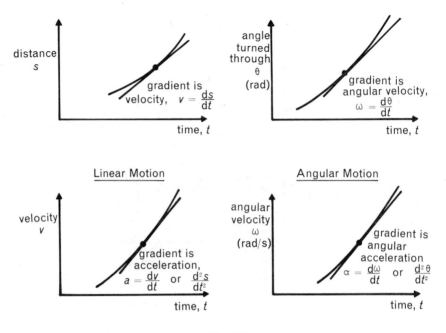

Fig. 2.2

EXAMPLE 6

A body moves a distance s metres in a time of t seconds so that
$s = 2t^3 - 9t^2 + 12t + 6$.

Find (a)　its velocity after 3 seconds,

(b)　its acceleration after 3 seconds, and

(c)　when the velocity is zero.

We have,
$$s = 2t^3 - 9t^2 + 12t + 6$$

$\therefore$
$$\frac{ds}{dt} = 6t^2 - 18t + 12$$

and
$$\frac{d^2s}{dt^2} = 12t - 18$$

(a)　When　$t = 3$　then velocity is　$\dfrac{ds}{dt} = 6(3)^2 - 18(3) + 12 = 12$ m/s

(b)　when　$t = 3$　then acceleration is　$\dfrac{d^2s}{dt^2} = 12(3) - 18 = 18$ m/s^2

(c)　when the velocity is zero then　$\dfrac{ds}{dt} = 0$

that is $\qquad 6t^2-18t+12 = 0$

$\therefore \qquad\qquad\qquad t^2-3t+2 = 0$

$\therefore \qquad\qquad\qquad (t-1)(t-2) = 0$

$\therefore$ either $\qquad\quad t-1 = 0 \quad$ or $\quad t-2 = 0$

$\therefore$ either $\qquad\quad t = 1$ second or $t = 2$ seconds.

EXAMPLE 7

The angle θ radians is connected with the time t seconds by the relationship $\theta = 20+5t^2-t^3$.

Find (a)　the angular velocity when　$t = 2$ seconds,　and

　　　(b)　the value of t when the angular deceleration is 4 rad/s².

We have, $\qquad\qquad\qquad \theta = 20+5t^2-t^3$

$\therefore \qquad\qquad\qquad\qquad \dfrac{d\theta}{dt} = 10t-3t^2$

and $\qquad\qquad\qquad\qquad \dfrac{d^2\theta}{dt^2} = 10-6t$

(a)　when　$t = 2$,　then the angular velocity $\dfrac{d\theta}{dt} = 10(2)-3(2)^2 = 8$ rad/s.

(b)　an angular deceleration of 4 rad/s² may be called an angular acceleration of -4 rad/s²,

$\therefore$ when $\qquad\quad \dfrac{d^2\theta}{dt^2} = -4 \quad$ then $\quad -4 = 10-6t$

$\qquad\qquad\qquad\qquad$ or $\qquad t = 2.33$ seconds.

Exercise 6

1)　If　$s = 10+50t-2t^2$,　where s metres is the distance travelled in t seconds by a body, what is the velocity of the body after 2 seconds?

2)　If　$v = 5+24t-3t^2$ where v m/s is the velocity of a body at a time t seconds, what is the acceleration when　$t = 3$?

3)　A body moves s metres in t seconds where　$s = t^3-3t^2-3t+8$.

Find　(a)　its velocity at the end of 3 seconds,

　　　(b)　when its velocity is zero,

　　　(c)　its acceleration at the end of 2 seconds,

　　　(d)　when its acceleration is zero.

4) A body moves s metres in t seconds where $s = \dfrac{1}{t^2}$. Find the velocity and acceleration after 3 seconds.

5) The distance s metres travelled by a falling body starting from rest after a time t seconds is given by $s = 5t^2$. Find its velocity after 1 second and after 3 seconds.

6) The distance s metres moved by the end of a lever after a time t seconds is given by the formula $s = 6t^2$. Find the velocity of the end of the lever when it has moved a distance $\frac{1}{2}$ metre.

7) The angular displacement θ radians of the spoke of a wheel is given by the expression $\theta = \dfrac{1}{2}t^4 - t^3$ where t seconds is the time.

Find   (a)   the angular velocity after 2 seconds,

(b)   the angular acceleration after 3 seconds,

(c)   when the angular acceleration is zero.

8) An angular displacement θ radians in time t seconds is given by the equation $\theta = \sin 3t$.

Find   (a)   the angular velocity when $t = 1$ second,

(b)   the smallest positive value of t for which the angular velocity is 2 rad/s,

(c)   the angular acceleration when $t = 0.5$ seconds,

(d)   the smallest positive value of t for which the angular acceleration is 9 rad/s².

9) A mass of 5000 kg moves along a straight line so that the distance s metres travelled in a time t seconds is given by $s = 3t^2 + 2t + 3$. If v m/s is its velocity and m kg is its mass then its kinetic energy is given by the formula $\dfrac{1}{2}mv^2$. Find its kinetic energy at a time $t = 0.5$ seconds remembering that the joule (J) is the unit of energy.

MAXIMUM AND MINIMUM

On reaching the end of this chapter you should be able to :-

1. *Define the turning point of a graph.*
2. *Determine the derivative of the function of the graph concerned.*
3. *Determine the value of x (the independent variable) at the turning points using 1 and 2.*
4. *Evaluate y (the dependent variable) corresponding to the values in 3.*
5. *Determine the nature of the turning points by consideration of the gradient on either side of the point.*
6. *Determine and evaluate the second derivative of the function at the turning points.*
7. *Determine the nature of the turning points by the sign of the second derivative.*
8. *Solve problems involving maxima and minima relevant to technology.*

TURNING POINTS

At the points P and Q (Fig. 3.1) the tangent to the curve is parallel to the *x*-axis. The points P and Q are called *turning points*. The turning point at P is called a *maximum* turning point and the turning point at Q is called a *minimum* turning point. It will be seen from Fig. 3.1 that the value of *y* at P is not the greatest value of *y* nor is the value of *y* at Q the least. The terms maximum and minimum values apply only to the values of *y* at the turning points and not to the values of *y* in general.

In practical applications, however, we are usually concerned with a specific range of values of *x* which are dictated by the problem. There is then no difficulty in identifying a particular maximum or minimum within this range of values of *x*.

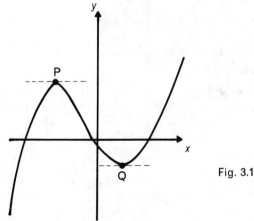

Fig. 3.1

EXAMPLE 1

Plot the graph of $y = x^3 - 5x^2 + 2x + 8$ for values of x between -2 and 6. Hence find the maximum and minimum values of y.

To plot the graph we draw up a table in the usual way.

x	-2	-1	0	1	2	3	4	5	6
$y = x^3 - 5x^2 + 2x + 8$	-24	0	8	6	0	-4	0	18	56

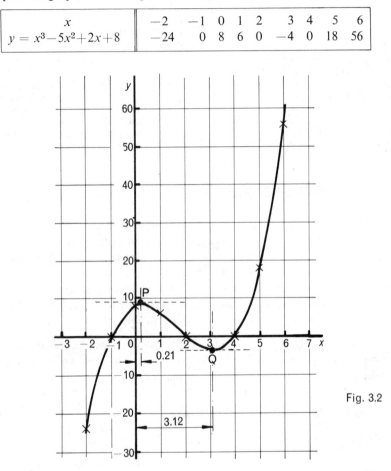

Fig. 3.2

The graph is shown in Fig. 3.2. The maximum value occurs at the point P where the tangent to the curve is parallel to the x-axis. The minimum value occurs at the point Q where again the tangent to the curve is parallel to the x-axis. From the graph the maximum value of y is 8.21 and the minimum value of y is -4.06.

Notice that the value of y at P is not the greatest value of y nor is the value of y at Q the least. However, the values of y at P and Q are called the maximum and minimum values of y respectively.

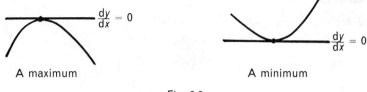

Fig. 3.3

It is not always convenient to draw the full graph to find the turning points as in the previous example. At a turning point the tangent to the curve is parallel to the x-axis (Fig. 3.3) and hence the gradient of the curve is zero, i.e. $\dfrac{\mathrm{d}y}{\mathrm{d}x} = 0$. Using this fact enables us to find the values of x at which the turning points occur.

It is then necessary to determine whether the points are maximum or minimum.

Two methods of testing are as follows:

Method 1

Consider the gradients of the curve on either side of the turning point. Fig. 3.4 shows how the gradient (or slope) of curve changes in the vicinity of a turning point.

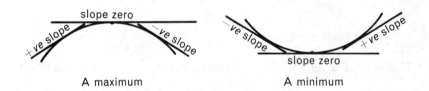

Fig. 3.4

Method 2

Find the value of $\dfrac{\mathrm{d}^2 y}{\mathrm{d}x^2}$ at the turning point.

If it is positive then the turning point is a minimum, and if it is negative then the turning point is a maximum.

If the original expression may be differentiated twice and the expression for $\dfrac{\mathrm{d}^2 y}{\mathrm{d}x^2}$ obtained without too much difficulty, then the second method is generally used.

EXAMPLE 2

Find the maximum and minimum values of y given that $y = x^3 + 3x^2 - 9x + 6$.

We have,

$$y = x^3 + 3x^2 - 9x + 6$$

$\therefore$

$$\frac{dy}{dx} = 3x^2 + 6x - 9$$

and

$$\frac{d^2y}{dx^2} = 6x + 6$$

At a turning point

$$\frac{dy}{dx} = 0$$

$\therefore$ $$3x^2 + 6x - 9 = 0$$

$\therefore$ $$x^2 + 2x - 3 = 0 \quad \text{by dividing through by 3.}$$

$\therefore$ $$(x-1)(x+3) = 0$$

$\therefore$ either $x - 1 = 0$ or $x + 3 = 0$

$\therefore$ either $x = 1$ or $x = -3$

Test for maximum or minimum:

From above we have $$\frac{d^2y}{dx^2} = 6x + 6$$

$\therefore$ at the point where $x = 1$, $\dfrac{d^2y}{dx^2} = 6(1) + 6 = +12$

This is positive and hence the turning point at $x = 1$ is a minimum.

The minimum value of y may be found by substituting $x = 1$ into the given equation. Hence:

$$y_{min} = (1)^3 + 3(1)^2 - 9(1) + 6 = 1.$$

At the point where $x = -3$, $\dfrac{d^2y}{dx^2} = 6(-3) + 6 = -12.$

This is negative and hence at $x = -3$ there is a maximum turning point. The maximum value of y may be found by substituting $x = -3$ into the given equation. Hence:

$$y_{max} = (-3)^3 + 3(-3)^2 - 9(-3) + 6 = +33.$$

To illustrate the test for maximum and minimum using the tangent gradient method, this method will be used to verify the above results.

At the turning point where $x = 1$, we know that,

$$\frac{dy}{dx} = 0, \quad \text{i.e. there is zero slope,}$$

and using a value of x slightly less than 1, say $x = 0.5$ then,

$$\frac{dy}{dx} = 3(0.5)^2 + 6(0.5) - 9 = -5.25 \quad \text{i.e. there is a negative slope,}$$

and using a value of x slightly greater than 1, say $x = 1.5$ then,

$$\frac{dy}{dx} = 3(1.5)^2 + 6(1.5) - 9 = +6.75 \quad \text{i.e. there is a positive slope.}$$

These results are best shown by means of a diagram (Fig. 3.5) which indicates clearly that when $x = 1$ we have a minimum.

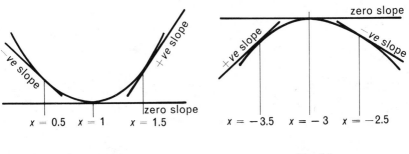

Fig. 3.5   Fig. 3.6

Now at the turning point where $x = -3$ we know that,

$$\frac{dy}{dx} = 0, \quad \text{i.e. there is zero slope,}$$

and using a value of x slightly less than -3, say $x = -3.5$ then,

$$\frac{dy}{dx} = 3(-3.5)^2 + 6(-3.5) - 9 = +6.75 \quad \text{i.e. there is a positive slope,}$$

and using a value of x slightly greater than -3, say $x = -2.5$ then,

$$\frac{dy}{dx} = 3(-2.5)^2 + 6(-2.5) - 9 = -5.25 \quad \text{i.e. there is a negative slope.}$$

Fig. 3.6 indicates that when $x = -3$ we have a maximum turning point.

There are many applications in engineering which involve the finding of maxima and minima. The first step is to construct an equation connecting the quantity for which a maximum or minimum is required in terms of another variable. A diagram representing the problem may help in the formation of this initial equation.

EXAMPLE 3

A rectangular sheet of metal 360 mm by 240 mm has four equal squares cut out at the corners. The sides are then turned up to form a rectangular box. Find the length of the sides of the squares cut out so that the volume of the box may be as great as possible, and find this maximum volume.

Let the length of the side of each cut away square be x mm as shown in Fig. 3.7.

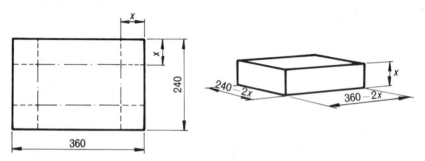

Fig. 3.7

Hence the volume is
$$V = x(240-2x)(360-2x)$$
$$= 4x^3 - 1200x^2 + 86\ 400x$$

$\therefore$
$$\frac{\mathrm{d}V}{\mathrm{d}x} = 12x^2 - 2400x + 86\ 400$$

and
$$\frac{\mathrm{d}^2V}{\mathrm{d}x^2} = 24x - 2400$$

At a turning point
$$\frac{\mathrm{d}V}{\mathrm{d}x} = 0$$

$\therefore \qquad 12x^2 - 2400x + 86\ 400 = 0$

or $\qquad x^2 - 200x + 7200 = 0$ by dividing through by 12.

Now this is a quadratic equation which does not factorise so we will have to solve using the formula for the standard quadratic $ax^2 + bx + c = 0$ which gives $x = \dfrac{-b \pm \sqrt{b^2 - 4ac}}{2a}$. Hence the solution of our equation is $\qquad x = \dfrac{-(-20)\pm\sqrt{(-20)^2 - 4\times 1\times 72}}{2\times 1}$

$\therefore$ either $\qquad x = 152.9 \quad$ or $\quad x = 47.1$

However, from the physical sizes of the sheet, it is not possible for x to be 152.9 mm (since one side is only 240 mm long) so we reject this solution. Hence $x = 47.1$ mm.

Test for maximum or minimum:

From the above we have $\qquad \dfrac{\mathrm{d}^2 V}{\mathrm{d}x^2} = 24x - 2400$

and hence when $\quad x = 47.1, \quad \dfrac{\mathrm{d}^2 V}{\mathrm{d}x^2} = 24(47.1) - 2400 = -1270$

This is negative and hence V is a maximum when $\quad x = 47.1$ mm.

It only remains to find the maximum volume by substituting $\quad x = 47.1$ into the equation for V.

$\therefore \qquad V_{\text{max}} = 47.1(240 - 2\times 47.1)(360 - 2\times 47.1) = 1.825 \times 10^6 \text{ mm}^3.$

EXAMPLE 4

A cylinder with an open top has a capacity of 2 m³ and is made from sheet metal. Neglecting any overlaps at the joints find the dimensions of the cylinder so that the amount of sheet steel used is a minimum.

Let the height of the cylinder be h metres and the radius of the base be r metres as shown in Fig. 3.8.

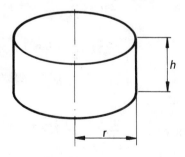

Fig. 3.8

Now the total area of metal = area of base+area of curved side:

∴ $$A = \pi r^2 + 2\pi r h$$

We cannot proceed to differentiate as there are two variables on the right hand side of the equation. It is possible, however, to find a connection between r and h using the fact that the volume is 2 m².

Now   volume of a cylinder $= \pi r^2 h$

∴ $$2 = \pi r^2 h$$

from which $$h = \frac{2}{\pi r^2}$$

We may now substitute for h is the equation for A,

∴ $$A = \pi r^2 + 2\pi r\left(\frac{2}{\pi r^2}\right)$$

∴ $$= \pi r^2 + \frac{4}{r}$$

∴ $$= \pi r^2 + 4r^{-1}$$

∴ $$\frac{dA}{dr} = 2\pi r - 4r^{-2}$$

and $$\frac{d^2A}{dr^2} = 2\pi + 8r^{-3}$$

Now for a turning point $$\frac{dA}{dr} = 0$$

or $$2\pi r - 4r^{-2} = 0$$

∴ $$2\pi r - \frac{4}{r^2} = 0$$

∴ $$2\pi r = \frac{4}{r^2}$$

∴ $$r^3 = \frac{2}{\pi} = 0.637$$

$$r = \sqrt[3]{0.637} = 0.860$$

To test for a minimum:

From above we have $$\frac{d^2A}{dr^2} = 2\pi + 8r^{-3}$$

$$= 2\pi + \frac{8}{r^3}$$

We do not need to do any further calculation here as this expression must be positive for all positive values of r. Hence $r = 0.86$ makes A a minimum.

We may find the corresponding value of h by substituting $r = 0.86$ into the equation found previously for h in terms of r

$$\therefore \qquad h = \frac{2}{\pi(0.86)^2} = 0.86$$

hence for the minimum amount of metal to be used the radius is 0.86 m and the height is 0.86 m.

Exercise 7

1) Find the maximum and minimum values of:

(a) $y = 2x^3 - 3x^2 - 12x + 4$

(b) $y = x^3 - 3x^2 + 4$

(c) $y = 6x^2 + x^3$

2) Given that $y = 60x + 3x^2 - 4x^3$, calculate:

(a) the gradient of the tangent to the curve of y at the point where $x = 1$;

(b) the value of x for which y has its maximum value;

(c) the value of x for which y has its minimum value.

3) Calculate the co-ordinates of the points on the curve $y = x^3 - 3x^2 - 9x + 12$ at each of which the tangent to the curve is parallel to the x-axis.

4) A curve has the equation $y = 8 + 2x - x^2$. Find:

(a) the value of x for which the gradient of the curve is 6;

(b) the value of x which gives the maximum value of y;

(c) the maximum value of y.

5) The curve $y = 2x^2 + \dfrac{k}{x}$ has a gradient of 5 when $x = 2$.

Calculate (a) the value of k; (b) the minimum value of y.

6) From a rectangular sheet of metal measuring 120 mm by 75 mm equal squares of side x are cut from each of the corners. The remaining flaps are then folded upwards to form an open box. Prove that the volume of the box is given by $V = 9000x - 390x^2 + 4x^3$. Find the value of x such that the volume is a maximum.

7) An open rectangular tank of height h metres with a square base of side x metres is to be constructed so that it has a capacity of 500 cubic metres. Prove that the surface area of the four walls and the base will be $\left(\dfrac{2000}{x}+x^2\right)$ square metres. Find the value of x for this expression to be a minimum.

8) The volume of a cone is given by the formula $V = \frac{1}{3}\pi r^2 h$, where h is the height of the cone and r its radius. If $h = 6-r$, calculate the value of r for which the volume is a maximum.

9) A box without a lid has a square base of side x mm and rectangular sides of height h mm. It is made from 10 800 mm^2 of sheet metal of negligible thickness. Prove that $h = \dfrac{10\,800-x^2}{4x}$ and that the volume of the box is $(2700x-\frac{1}{4}x^3)$. Hence calculate the maximum volume of the box.

10) A cylindrical tank, with an open top, is to be made to hold 300 cubic metres of liquid. Find the dimensions of the tank so that its surface area shall be a minimum.

11) A cooling tank is to be made with the trapezoidal section as shown:

Its cross-sectional area is to be 300 000 mm^2. Show that the width of material needed to form, from one sheet, the bottom and folded-up sides is $w = \dfrac{300\,000}{h}+1.828h$. Hence find the height h of the tank so that the width of material needed is a minimum.

12) A cylindrical cup is to be drawn from a disc of metal of 50 mm diameter. Assuming that the surface area of the cup is the same as that of the disc find the dimensions of the cup so that its volume is a maximum.

13) A lever weighing 12 N per m run of its length is as shown:

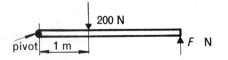

Find the length of the lever so that the force F shall be a minimum.

14) The cost per hour of running a certain machine is $C = 1.20 + 0.06\,N^3$ where N is the number of components produced per hour. Find the most economical value of N if 1000 components are to be produced.

(*Hint:* Establish the time taken to produce 1000 and then the total cost.)

15) A rectangle is inscribed in a circle of 120 mm diameter. Show that the rectangle having the largest area is a square, and find the length of its side.

16) The efficiency of a steam turbine is given by:

$$\eta = 4(n\rho \cos \alpha - n^2\rho^2)$$

where n and α are constants. Find the maximum value of η.

SUMMARY

Two tests for maximum or minimum:

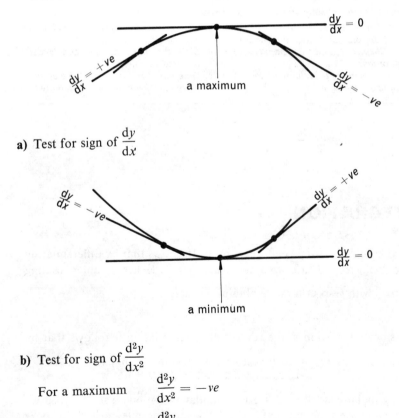

a) Test for sign of $\dfrac{dy}{dx}$

b) Test for sign of $\dfrac{d^2y}{dx^2}$

For a maximum $\dfrac{d^2y}{dx^2} = -ve$

and for a minimum $\dfrac{d^2y}{dx^2} = +ve$

AREAS AND VOLUMES BY INTEGRATION

On reaching the end of this chapter you should be able to :-

1. Determine the areas between the x-axis, the curve and given ordinates for functions in Objective 1, Chapter 1, and polynomials of degree three or less.

2. Sketch a typical incremental volume under the graph of a function listed in 1.

3. Determine the volume of the increment in 2 in a suitable notation, e.g. $\pi\, y^2 dx$.

4. State the limits of the variable defining the required volume.

5. Determine the required volume by summing the incremental volumes using definite integration.

INTEGRATION

The table of differential coefficients on p. 18 shows that by differentiating sin ax with respect to x we obtain $a.\cos ax$. Hence by differentiating $\frac{1}{a}.\sin ax$ with respect to x we obtain cos ax.

Now since integration is the reverse of differentiation it follows that by integrating cos ax with respect to x we obtain $\frac{1}{a}.\sin ax$.

Now by making small modifications similar to the one just described we may rewrite the table on p. 18 as a table showing integrals of the more common functions.

44

y	$\int y.dx$
$a.x^n$	$\dfrac{a}{n+1}.x^{n+1}$
$\sin ax$	$-\dfrac{1}{a}.\cos ax$
$\cos ax$	$\dfrac{1}{a}.\sin ax$
$\sec^2 x$	$\tan x$
$\dfrac{1}{x}$	$\log_e x$
e^{ax}	$\dfrac{1}{a}.e^{ax}$

The worked examples which follow will remind you of the integration process and also the use of the above table of integrals.

EXAMPLE 1

Find $\int (x^2 + 2x - 3)\,dx$

$$\int (x^2 + 2x - 3)\,dx = \frac{x^3}{3} + 2\frac{x^2}{2} - 3x + k$$

$$= \frac{x^3}{3} + x^2 - 3x + k$$

Remember that this is an 'indefinite integral', since there were no limits given, and the solution must contain a constant of integration, k.

EXAMPLE 2

Find the value of $\int_1^2 x^3 + 4\,dx$

$$\int_1^2 x^3 + 4\,dx = \left[\frac{x^4}{4} + 4x\right]_1^2$$

$$= \left(\frac{2^4}{4} + 4 \times 2\right) - \left(\frac{1^4}{4} + 4 \times 1\right)$$

$$= 4 + 8 - 0.25 - 4$$

$$= 7.75$$

Since we were given limits this integral is called a 'definite integral'. Note the use of the square brackets which mean 'we have completed the integration and will next be substituting the limits'. Also in the evaluation of a definite integral there is no need to introduce a constant of integration.

EXAMPLE 3

Evaluate $\displaystyle\int_0^2 \sin 2\theta \, d\theta$

$$\int_0^2 \sin 2\theta \, d\theta = \left[-\frac{1}{2} \cos 2\theta \right]_0^2$$

$$= \left\{ -\frac{1}{2} \cos (2 \times 2) \right\} - \left\{ -\frac{1}{2} \cos (2 \times 0) \right\}$$

$$= -\frac{1}{2} \cos 4 + \frac{1}{2} \cos 0$$

$$= 0.327 + 0.5$$

$$= 0.827$$

In this example the values of the angles, after substituting the limits, are in radians.

Exercise 8

1) Find $\displaystyle\int \left(5x^3 + 2x - \frac{4}{x^2} \right) dx$ **2)** Find $\displaystyle\int \left(\sqrt{x} + \frac{1}{\sqrt{x}} \right) dx$

Evaluate the definite integrals in Questions 3 to 10:

3) $\displaystyle\int_0^1 \cos \theta \, d\theta$ **7)** $\displaystyle\int_0^\pi \sec^2 \phi \, d\phi$

4) $\displaystyle\int_0^{\pi/2} \sin \theta \, d\theta$ **8)** $\displaystyle\int_{-1}^1 e^{2t} \, dt$

5) $\displaystyle\int_2^3 e^{-t} \, dt$ **9)** $\displaystyle\int_0^2 \frac{x^2 + x^3}{x} \, dx$

6) $\displaystyle\int_1^4 x - \frac{1}{x} \, dx$ **10)** $\displaystyle\int_{-\pi/2}^{\pi/2} \sin \tfrac{1}{2} t \, dt$

AREAS BY INTEGRATION

You will remember that we may apply integration to the finding of areas by considering the area to be the sum of a number of elementary strips.

EXAMPLE 4

Find the area enclosed by the curve $y = x^3 - 2x^2 + 5$, the x-axis, and the lines $x = 1$ and $x = 3$.

When applying integration to the finding of areas, volumes etc. it helps, whenever possible, to make a sketch of the problem. Fig. 4.1 shows a sketch of the required area and a typical elementary strip whose width δx is very small. Hence the elementary strip may be considered to be a rectangle having an area equal approximately to $y.\delta x$. The required area may be considered to be the sum of all such elementary strip areas between the values of $x = 1$ and $x = 3$.

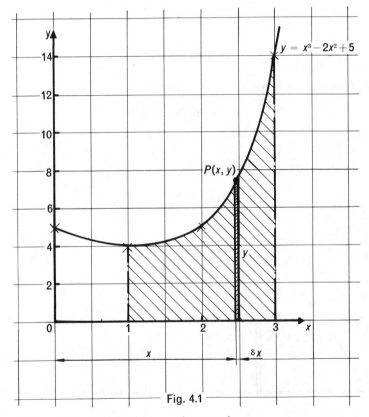

Fig. 4.1

In mathematical notation this may be stated as:

$$\text{Area} = \sum_{x=1}^{x=3} y.\delta x \qquad \text{approximately.}$$

The process of integration may be considered to sum up an infinite number of elementary strips and hence gives an exact result.

$\therefore$ Area $= \displaystyle\int_{1}^{3} y.dx$

$= \displaystyle\int_{1}^{3} (x^3 - 2x^2 + 5)\, dx$

$= \left[\dfrac{1}{4}x^4 - \dfrac{2}{3}x^3 + 5x \right]_{1}^{3}$

$= \left(\dfrac{1}{4} \times 3^4 - \dfrac{2}{3} \times 3^3 + 5 \times 3 \right) - \left(\dfrac{1}{4} \times 1^4 - \dfrac{2}{3} \times 1^3 + 5 \times 1 \right)$

$= 20.25 - 18 + 15 - 0.25 + 0.67 - 5$

$= 12.67$ square units.

EXAMPLE 5

Find the area under the curve $\ y = 4\sin 2\theta\ $ over a half-cycle.

Fig. 4.2 shows a sketch of the curve $\ y = 4\sin 2\theta\ $ over a half-cycle.
The limits of the integral will therefore be $\ \theta = 0\ $ and $\ \theta = \dfrac{\pi}{2}\ $ radians.

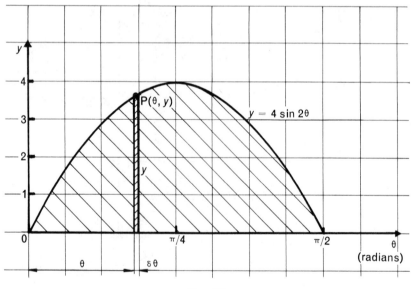

Fig. 4.2

As in the previous example:

required area $= \displaystyle\int y\, d\theta$

$$= \int_0^{\frac{\pi}{2}} 4 \sin 2\theta \ d\theta$$

$$= \left[-2 \cos 2\theta \right]_0^{\frac{\pi}{2}}$$

$$= \left\{ -2 \cos 2\left(\frac{\pi}{2}\right) \right\} - \left\{ -2 \cos 2(0) \right\}$$

$$= -2 \cos \pi + 2 \cos 0$$

$$= -2 \cos 180° + 2 \cos 0°$$

$$= -2(-1) + 2(1)$$

$$= 4 \text{ square units.}$$

EXAMPLE 6

Find the area between the curve $y = e^{-0.6t}$, the two axes and the line $t = 3$.

Fig. 4.3 shows a sketch of the curve and the required area.

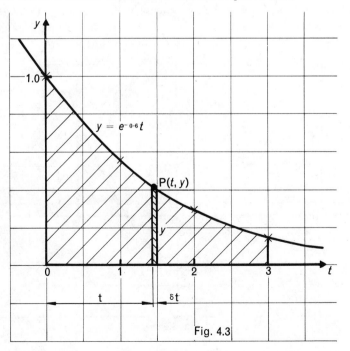

Fig. 4.3

As previously:

$$\text{area} = \int y \, dt$$

$$= \int_0^3 e^{-0.6t} \, dt$$

$$= \left[-\frac{1}{0.6} e^{-0.6t} \right]_0^3$$

$$= -\frac{1}{0.6} \left[e^{-0.6t} \right]_0^3$$

$$= -\frac{1}{0.6} (e^{-0.6 \times 3} - e^{-0.6 \times 0})$$

$$= -\frac{1}{0.6} (0.165 - 1.000)$$

$$= 1.392 \text{ square units.}$$

EXAMPLE 7

Sketch the curve $y = (x-1)(x+2)$ and hence find the area under the curve $y = (x-1)(x+2)$ between $x = -3$ and $x = 2$ and interpret the result.

The curve is sketched in Fig. 4.4 and it will be seen that it cuts the x-axis at $x = -2$ and $x = 1$. The area consists of three distinct parts as shown in the diagram, and each area will be calculated separately.

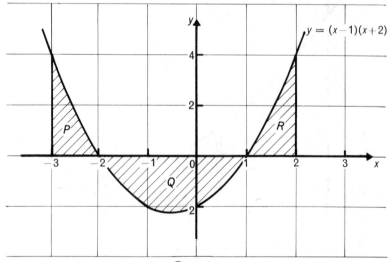

Fig. 4.4

Area $P = \int_{-3}^{-2} (x^2+x-2)\,dx = \left[\dfrac{x^3}{3}+\dfrac{x^2}{2}-2x\right]_{-3}^{-2}$

$= \left(\dfrac{(-2)^3}{3}+\dfrac{(-2)^2}{2}-2(-2)\right)-\left(\dfrac{(-3)^3}{3}+\dfrac{(-3)^2}{2}-2(-3)\right)$

$= 1.833$ square units

Area $Q = \int_{-2}^{1} (x^2+x-2)\,dx = \left[\dfrac{x^3}{3}+\dfrac{x^2}{2}-2x\right]_{-2}^{1}$

$= \left(\dfrac{(1)^3}{3}+\dfrac{(1)^2}{2}-2(1)\right)-\left(\dfrac{(-2)^3}{3}+\dfrac{(-2)^2}{2}-2(-2)\right)$

$= -4.500$ square units

Area $R = \int_{1}^{2} (x^2+x-2)\,dx = \left[\dfrac{x^3}{3}+\dfrac{x^2}{2}-2x\right]_{1}^{2}$

$= \left(\dfrac{(2)^3}{3}+\dfrac{(2)^2}{2}-2(2)\right)-\left(\dfrac{(1)^3}{3}+\dfrac{(1)^2}{2}-2(1)\right)$

$= 1.833$ square units

Note that the areas P and R which lie above the x-axis are positive areas whilst the area Q which lies below the x-axis is a negative area. If we integrate between the limits $x = -3$ and $x = 2$ we get:

$\int_{-3}^{2} (x^2+x-2)\,dx = \left[\dfrac{x^3}{3}+\dfrac{x^2}{2}-2x\right]_{-3}^{2}$

$= \left(\dfrac{(2)^3}{3}+\dfrac{(2)^2}{2}-2(2)\right)-\left(\dfrac{(-3)^3}{3}+\dfrac{(-3)^2}{2}-2(-3)\right)$

$= -0.833$ square units

Now   area P+area Q+area $R = 1.833-4.500+1.833$
$= -0.833$ square units.

Hence when we integrate between the limits $x = -3$ and $x = 2$, the area given by the integration is the *net* area, that is, the signs of the various areas are taken into account.

Before attempting to find the area under a curve by integration it pays to first sketch the curve.

Exercise 9

1) Find the area under the curve $y = x^2+3$ between $x = 1$ and $x = 3$.

2) Find the area under the curve $y = x^2-2x+2$ between $x = 0$ and $x = 2$.

3)   Find the area between the curve $y = (x-1)(x-2)$ and the x-axis.

4)   Find the area bounded by the curve $y = (1+x)(2-x)$ and the x-axis.

5)   Sketch the curve $y = x(x-1)(x-2)$ and find the two areas between the curve and the x-axis.

6)   Show that $\displaystyle\int_{-2}^{2} x(x^2-4)\,dx = 0$,   and interpret the result graphically.

7)   Find the area enclosed between the curve $y^2 = 4x$ and the line $y = x$.

8)   Find the area between the curve $y = \dfrac{1}{x}+x$,   the x-axis and the lines $x = 1$ and $x = 2$.

9)   Show that $\displaystyle\int_{0}^{2\pi} \sin x\,dx = 0$ and explain the result with reference to a graph.

10)   Find the area bounded by the curve $y = \cos x$,   the x-axis, and the lines $x = 0$ and $x = \dfrac{\pi}{3}$.

11)   Find the area under the curve $y = e^x$ between $x = -2$ and $x = +2$.

12)   Find the area under the curve $y = e^{2x}$ between $x = -2$ and $x = +1$.

13)   Find the area enclosed by the curve $y = e^{-t/2}$,   the two axes, and the line $t = 2$.

VOLUMES OF REVOLUTION BY INTEGRATION

If the area under the curve APB (Fig. 4.5) is rotated one complete revolution about the x-axis, then the volume swept out is called a volume of revolution.

The point P, whose co-ordinates are (x, y) is a point on the curve AB.

Let us consider, below P, a thin slice whose width is δx. Since the width of the slice is very small we may consider the slice to be a cylinder of radius y. The volume of this slice is approximately $\pi y^2.\delta x$.   Such a slice is called an elementary slice and we will consider that the volume of revolution is made up from many such elementary slices. Hence the complete volume of revolution is the sum of all the elementary slices between the values of $x = a$ and $x = b$.

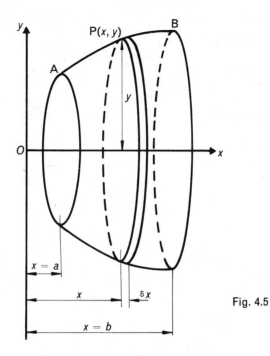

Fig. 4.5

In mathematical notation this may be stated as

$$\sum_{x=b}^{x=a} \pi y^2 . \delta x \quad \text{approximately.}$$

As for areas, the process of integration may be considered to sum an infinite number of elementary slices and hence it gives an exact result.

$\therefore$ $\quad$ Volume of revolution $= \displaystyle\int_a^b \pi y^2 \, \mathrm{d}x \quad$ exactly

EXAMPLE 8

The area between the curve $y = x^2$, the x-axis and the ordinates $x = 1$ and $x = 3$ is rotated about the x-axis. Find the volume of revolution.

As when finding areas it is recommended that a sketch is made of the required volume. Fig. 4.6 shows a sketch of the required volume.

Required volume of revolution $= \displaystyle\int_1^3 \pi y^2 \, \mathrm{d}x$

$$= \int_1^3 \pi (x^2)^2 \, \mathrm{d}x$$

$$= \pi \int_1^3 x^4 \, dx$$

$$= \pi \left[\frac{x^5}{5} \right]_1^3$$

$$= \pi \left\{ \frac{3^5}{5} - \frac{1^5}{5} \right\}$$

$$= 152.1 \text{ cubic units.}$$

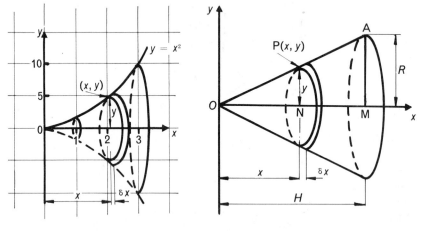

Fig. 4.6   Fig. 4.7

EXAMPLE 9

Find, by the calculus, the volume of a cone of base radius R and height H.

The first step is to set up the cone on suitable axes. For convienence the cone has been put with its polar axis lying along the x-axis as shown in Fig. 4.7.

Then the volume of the cone is the volume of revolution generated when the area OAM is rotated about the x-axis.

We need an equation connecting x and y. This may be found by considering the similar triangles OPN and OAM.

Then

$$\frac{PN}{ON} = \frac{AM}{OM}$$

or

$$\frac{y}{x} = \frac{R}{H}$$

or

$$y = \frac{R}{H} \cdot x$$

Now the required volume $= \displaystyle\int_0^H \pi y^2 \, \mathrm{d}x$

$$= \int_0^H \pi \left(\frac{R}{H}.x\right)^2 \mathrm{d}x$$

$$= \pi \frac{R^2}{H^2} \int_0^H x^2 \, \mathrm{d}x$$

$$= \pi \frac{R^2}{H^2} \left[\frac{x^3}{3}\right]_0^H$$

$$= \pi \frac{R^2}{H^2} \left\{\frac{H^3}{3} - \frac{0^3}{3}\right\}$$

$$= \frac{1}{3} \pi R^2 H$$

which verifies a formula you know already.

EXAMPLE 10

Find the volume of a sphere of radius R.

The volume of a sphere may be considered to be a volume of revolution generated by the revolution of a semi-circular area about its boundary diameter. Suitable axes must now be chosen and these are shown in Fig. 4.8.

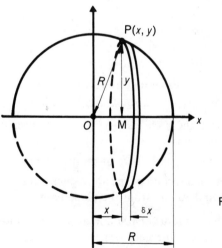

Fig. 4.8

The relationship connecting y and x may be found by considering the right-angled triangle OPM and applying the theorem of Pythagoras.

Hence $\qquad\qquad\qquad x^2+y^2 = R^2$

$\therefore \qquad\qquad\qquad y^2 = R^2-x^2$

Now the volume of revolution $= \displaystyle\int \pi y^2 \, dx$

$$= \pi \int_{-R}^{R} (R^2-x^2) \, dx$$

$$= \pi \left[R^2x-\frac{x^3}{3} \right]_{-R}^{R}$$

$$= \pi\left\{ \left(R^2(R)-\frac{R^3}{3} \right)-\left(R^2(-R)-\frac{(-R)^3}{3} \right)\right\}$$

$$= \frac{4}{3}\pi R^3$$

verifying a formula you should recognise.

Exercise 10

In Questions 1–5 find the volume generated about the x-axis of the given curves between the limits stated:

1) $y = x^3$ from $x = 0$ to $x = 2$.

2) $xy = 16$ from $x = 1$ to $x = 2$.

3) $x = 4\sqrt{y}$ from $x = 1$ to $x = 2$.

4) $y^3 = x^2$ from $x = 1$ to $x = 8$.

5) $y = \dfrac{x}{2}$ from $x = 0$ to $x = 4$.

6) Find the volume generated by revolving about the x-axis the portion of the curve $y = x(x-1)$ which lies below the x-axis.

7) The portion of the circle $x^2+y^2 = 25$ in the first quadrant is rotated about the x-axis. Find:

(a) the volume of a hemisphere of radius 50 mm,

(b) the volume of a section of the above hemi-sphere cut off by the planes distant 20 mm and 30 mm from the plane base.

8) A bucket has a radius of 100 mm at the base, and 200 mm at the top. It is 200 mm deep and the sides slope uniformly. Show that it may be considered as being formed by the revolution of the line $y = \dfrac{x}{2}+100$ about the x-axis from $x = 0$ to $x = 200$. Hence find the capacity of the bucket, when full, in litres.

9) Fig. 4.9 shows the cross-section through a brass nozzle. Find the volume of the nozzle.

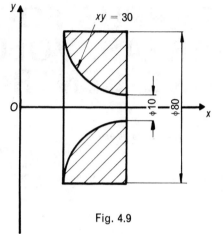

Fig. 4.9

SUMMARY

a) The table shows integrals of the more common functions:

y	$y\,dx$
$a.x^n$	$\dfrac{a}{n+1}.x^{n+1}$
$\sin ax$	$-\dfrac{1}{a}.\cos ax$
$\cos ax$	$\dfrac{1}{a}.\sin ax$
$\sec^2 x$	$\tan x$
$\dfrac{1}{x}$	$\log_e x$
e^{ax}	$\dfrac{1}{a}.e^{ax}$

b) Area under a curve $= \displaystyle\int_a^b y\,dx.$

c) Volume of solid of revolution $= \pi \displaystyle\int_a^b y^2\,dx.$

THEOREMS OF PAPPUS FOR AREAS, VOLUMES AND CENTROIDS

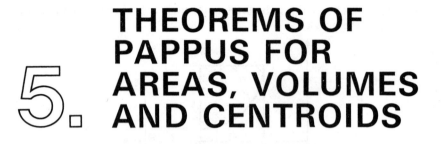

On reaching the end of this chapter you should be able to :-

1. Use Theorem 1 of Pappus to find volume, area, and position of centroid.

2. Apply 1 to determine the centroid of a semi-circular area.

3. Use Theorm 2 of Pappus to find area, length of arc, and position of centroid.

4. Apply the above Theorems to problems in engineering.

VOLUMES OF SOLIDS OF REVOLUTION

Theorem 1

If a plane area rotates about a line in its plane (which does not cut the area) then the volume generated is given by the equation:

Volume = area × length of path of its centroid

EXAMPLE 1

Find the volume of the circular ring of metal shown in Fig. 5.1.

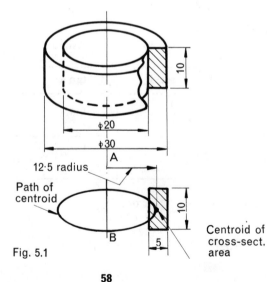

Fig. 5.1

Consider that the volume of the ring is made by one revolution of the cross-sectional area shown about the axis AB. Pappus' Theorem 1 states:

$$\text{Volume} = \text{area} \times \text{length of path of centroid of area}$$

Hence, for one revolution of the area about AB, we have:

$$\text{volume} = (10 \times 5) \times (2\pi \times 12.5)$$

$$= 1250\pi$$

$$= 3930 \text{ mm}^3$$

This can be checked by considering the ring as shown in Fig. 5.2.

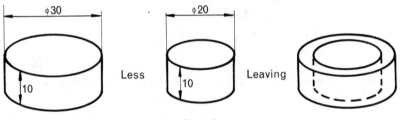

Fig. 5.2

We see that:

$$\text{vol. of ring} = (\text{vol. of 30 mm dia. disc}) - (\text{vol. of 20 mm dia. disc})$$

$$= (\pi 15^2 - \pi 10^2) \times 10$$

$$= 1250\pi$$

$$= 3930 \text{ mm}^3$$

which verifies the previous result.

EXAMPLE 2

Find the volume of a right circular cone having base radius a and height h.

By Pappus' Theorem 1 (see Fig. 5.3):

$$\text{volume} = \text{area} \times \text{path of centroid}$$

$$= \tfrac{1}{2}ha \times 2\pi \frac{a}{3}$$

$$= \tfrac{1}{3}\pi a^2 h$$

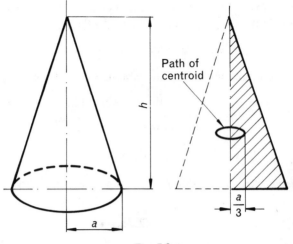

Fig. 5.3

You will recognise this as the correct formula for the volume of a right circular cone.

EXAMPLE 3

By rotating a semi-circular area, of radius a, about its boundary diameter find the distance of the centroid of the area from the diameter.

When a semi-circular area is rotated about its boundary diameter a sphere is generated. Since the radius of the semi-circle is a, the volume of the sphere generated is $4\pi a^3/3$.

Let the required distance be $\bar{x}$ in Fig. 5.4.

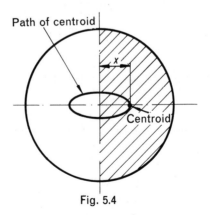

Fig. 5.4

By Pappus' Theorem 1,

$$\text{volume} = \text{area} \times \text{path of centroid of the area},$$

$$\therefore \qquad 4\pi a^3/3 = \tfrac{1}{2}\pi a^2 \times 2\pi \bar{x}$$

$$\therefore \qquad \bar{x} = \frac{4a}{3\pi} = 0.424a$$

This answer will prove useful in further work, and should be learnt.

EXAMPLE 4

A uniform solid circular cylinder of diameter 60 mm and height of 15 mm is to be made into a pulley wheel by cutting a groove round the curved surface. The cross-section of the groove is to be a semi-circle of diameter 10 mm. Find the mass of the pulley wheel if the density of the material is 8 g/cm³.

From the previous example the distance of the centroid of the semi-circular cross-sectional area of the groove from its diameter is $0.424 \times \dfrac{10}{2} = 2.12$ mm, so the distance of the centroid from the axis of the cylinder $= \dfrac{60}{2} - 2.12 = 27.88$ mm (see Fig. 5.5).

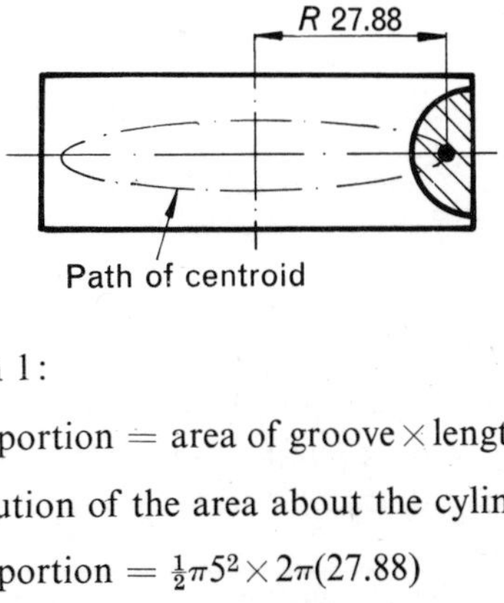

Fig. 5.5

Path of centroid

By Pappus' Theorem 1:

Volume of cut-away portion = area of groove × length of path of centroid

and so for one revolution of the area about the cylinder axis,

Volume of cut-away portion $= \tfrac{1}{2}\pi 5^2 \times 2\pi(27.88)$

$$= 6880 \text{ mm}^3$$

but

Volume of pulley = volume of cylinder − volume of cut-away portion

$$= \frac{\pi}{4}(60)^2(15) - 6880$$

$$= 42\,410 - 6880$$

$$= 35\,530 \text{ mm}^3$$

now

$$\text{Mass of pulley} = \text{volume of pulley} \times \text{density}$$

$$= 35\,530 \times \frac{8}{1000}$$

$$= 284 \text{ g}$$

EXAMPLE 5

Find the volume of the component shown in Fig. 5.6.

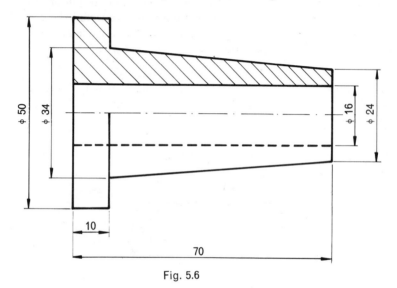

Fig. 5.6

The solid of revolution is obtained by rotating the shape shown in Fig. 5.7 about the axis XX. This shape can be conveniently divided into the three parts A, B and C which can then be dealt with separately as shown in the table below.

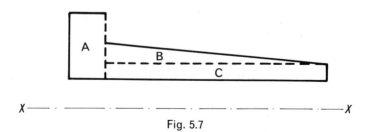

Fig. 5.7

Part	Area	$\bar{y}$	Area $\times 2\pi\bar{y}$
A	$17 \times 10 = 170$	16.5	$5\,610\pi$
B	$\frac{1}{2} \times 5 \times 60 = 150$	13.7	4110π
C	$4 \times 60 = 240$	10	$4\,800\pi$
		Volume $= 14\,520\pi$	

Volume of solid of revolution $= 14\,520\pi = 45\,600$ mm³

Exercise 11

1) By rotating the right angled triangle shown in Fig. 5.8 about its vertical side, find the volume of the cone generated.

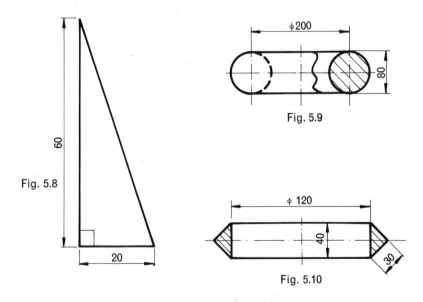

Fig. 5.9

Fig. 5.8

Fig. 5.10

2) Find the mass of the anchor ring shown in Fig. 5.9 if the density of the material is 7700 kg/m³.

3) Calculate the volume of the cast gunmetal ring shown in Fig. 5.10.

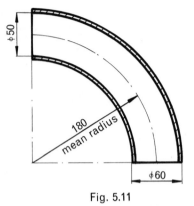

Fig. 5.11

4)   The right angled bend shown in Fig. 5.11 is formed from a steel pipe. Find its mass if the density of steel is 7800 kg/m³.

5)   A groove of width 20 mm and of semi-circular cross-section is turned on a metal rod of 50 mm diameter. Find the volume of metal removed.

6)   Find the volume of metal used in the casting shown in Fig. 5.12.

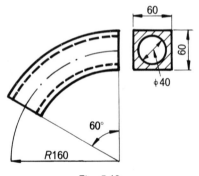

Fig. 5.12

7)   Find the volume of the turned component shown in Fig. 5.13:

(a)   using a theorem of Pappus,

(b)   by considering the shape as the frustum of a cone.

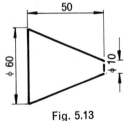

Fig. 5.13

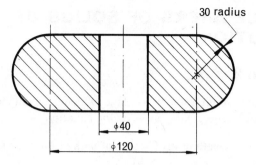

Fig. 5.14

8) Find the mass of the ring shown in Fig. 5.14 which is made in aluminium which has a density of 2700 kg/m³.

9) Fig. 5.15 shows a turned component. Find its volume.

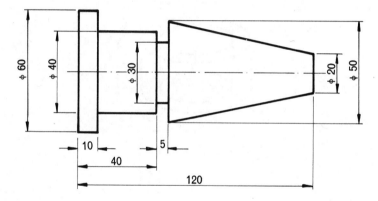

Fig. 5.15

10) Find the volume of the component shown in Fig. 5.16.

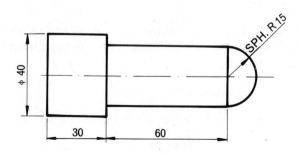

Fig. 5.16

SURFACE AREAS OF SOLIDS OF REVOLUTION

Theorem 2

If an arc rotates about a line in its plane (which does not cut the arc) then the surface area generated is given by the equation

Area = length of arc × length of path of the arc centroid

EXAMPLE 6

Find the surface area of the circular ring of metal shown in Fig. 5.17.

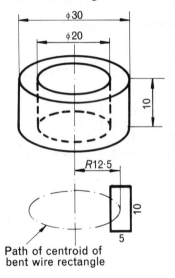

Fig. 5.17

Consider that the surface area of the ring is made by one revolution of a wire, bent to the shape of the perimeter of the rectangular cross-section, around the polar axis of the ring.

Pappus' Theorem 2 states that:

$$\text{area swept} = \text{length of arc} \times \text{path of centroid}$$

$$= (10+10+5+5) \times 2 \times 12.5 \times \pi$$

$$= 2350 \text{ mm}^2$$

EXAMPLE 7

Find the curved surface area of a right circular cone having base radius a and slant length l (see Fig. 5.18).

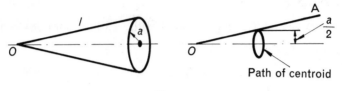

Fig. 5.18

Consider that the curved surface area of the cone is generated by one revolution of a straight length, l, of wire about the horizontal axis of the cone.

From Pappus' Theorem 2,

$$\text{surface area} = \text{length of line} \times \text{path of centroid}$$

so for one revolution of line OA about the axis OX we have:

$$\text{curved surface area} = l \times 2\pi\left(\frac{a}{2}\right)$$

$$= \pi a l$$

This verifies a result which you should know already.

EXAMPLE 8

By rotating a wire bent in the form of a semi-circular arc of radius R about a boundary diameter as shown in Fig. 5.19 find the distance of the centre of gravity of the wire from the diameter.

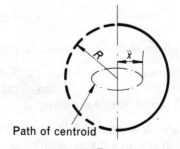

Path of centroid

Fig. 5.19

When a wire bent in the form of a semi-circular arc is rotated about its boundary diameter the surface of a sphere is generated. Since the radius of the semi-circle is R, the surface area generated is $4\pi R^2$.

Pappus' Theorem 2 states that:

$$\text{surface area generated} = \text{arc length} \times \text{path of centroid}$$

so for one revolution, if the required distance is $\bar{x}$,

$$\text{Surface area of sphere} = \pi R \times 2\pi \bar{x}$$

$$\therefore \qquad 4\pi R^2 = 2\pi^2 R\bar{x}$$

$$\therefore \qquad \bar{x} = \frac{2R}{\pi} = 0.637R$$

Exercise 12

1) By rotating the right angled triangle shown in Fig. 5.20 about its vertical side, find the curved surface area of the cone generated.

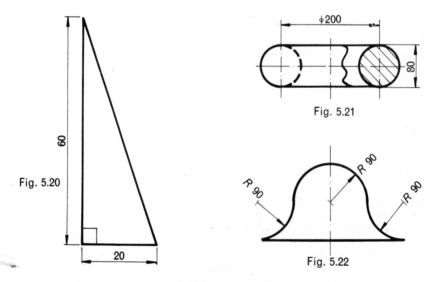

Fig. 5.21

Fig. 5.20

Fig. 5.22

2) Find the surface area of the anchor ring shown in Fig. 5.21.

3) By using the result obtained in worked Example 7, find the curved surface area of a groove of width 20 mm and of semi-circular cross-section which is turned on a metal rod of 50 mm diameter.

4) A spun cap has the shape shown in Fig. 5.22. Neglecting the thickness of the material find the total (i.e. inside and outside) surface area which will have to be painted.

5) Find the total surface area (inside and outside) of the casting shown in Fig. 23.

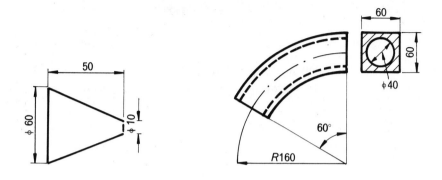

Fig. 5.23   Fig. 5.24

6) Find the curved surface area of the turned component shown in Fig. 5.24:

(a) using a theorem of Pappus,

(b) by considering the shape as frustum of a cone.

SUMMARY

a) Volume of solid of revolution $=$ area $\times$ length of path of centroid.

b) Surface area of solid of revolution $=$ length of arc $\times$ length of path of centroid.

c) The centroid of a semi-circular area is as shown in Fig. 5.25.

d) The centroid of the circumference of a semi-circle is shown in Fig. 5.26.

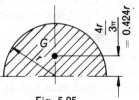

Fig. 5.25

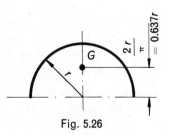

Fig. 5.26

CENTROIDS OF AREAS

On reaching the end of this chapter you should be able to :-

1. Sketch a given area including a typical incremental area whose centroid is known.
2. Determine the moment of the increment in 1 about a specified axis in the plane of the area.
3. Sum the moments of the increments between given limits by definite integration.
4. Determine the given area.

5. Define centroid as the ratio of the total moment determined in 3 to the total area determined in 4.
6. Calculate the distance of the centroid from the given axis.
7. Calculate the centroids of a rectangle, triangle, circle and sector of a circle.

CENTROIDS OF AREAS

We know that the moment of a force F about the line CD (Fig. 6.1) is given by $F \times d$.

Similarly we say that the first moment of the area A about the line CD (Fig. 6.2) is given by $A \times d$, the point G being the centroid of the area.

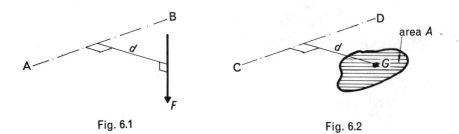

Fig. 6.1   Fig. 6.2

The centroid of an area is the point at which the total area may be considered to be situated for calculation purposes.

The centroid of an area is at the point which corresponds to the centre of gravity of a lamina of the same shape as the area. A thin flat sheet of steel of uniform thickness is an example of a lamina.

We know the position of the centres of gravity of the commonest shapes of laminae: for example the centre of gravity of a thin flat circular disc is at the centre of the disc. Hence we may say that the centroid of a circular area is at the centre of the area.

70

It is often possible to deduce the position of the centroid by the symmetry of the area. For example, in the case of a rectangle the centroid will be at the intersection of the centre-lines, as shown in Fig. 6.3, as the area is symmetrical about either of the centre-lines.

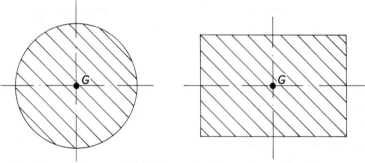

Positions of the centroid *G* for a circular and a rectangular area.

Fig. 6.3.

It is often necessary to find the position of the centroid of a *composite area*. This is an area which is made up from common shapes. To determine the position of the centroid we use two reference axes *Ox* and *Oy*. The location of the centroid is then given by the co-ordinates $\bar{x}$ and $\bar{y}$ as shown in Fig. 6.4.

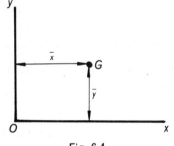

Fig. 6.4

To find $\bar{x}$ we use the formula $(\Sigma A)\bar{x} = \Sigma(Ax)$

where ΣA is the sum of the component areas

and ΣAx is the sum of the first moments of the component areas about the axis *Oy*.

To find $\bar{y}$ we use the formula $(\Sigma A)\bar{y} = (\Sigma Ay)$

You have used these formulae previously in *Level II Analytical Mathematics* and the following example will serve as a reminder.

EXAMPLE 1

Find the position of the centroid shown in Fig. 6.5.

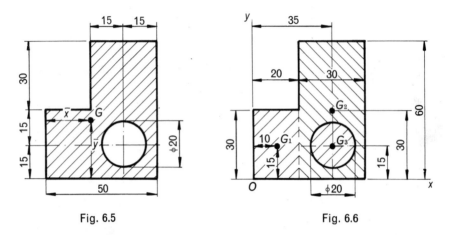

Fig. 6.5   Fig. 6.6

For convenience the reference axes Ox and Oy have been chosen as shown in Fig. 6.6. It will also be seen that the given area has been divided into three component areas whose centroids are G_1, G_2 and G_3. The dimensions of each component area and the location of each centroid from the axes Ox and Oy are clearly shown. A diagram of this type is essential for this and similar problems — any attempt to obtain these dimensions mentally usually results in one or more errors. Remember the circle is a 'missing' area and must be taken as negative in the calculations.

It helps to simplify the arithmetic and hence reduce errors by showing the solution in tabular form.

Area	Distance to centroid		Moment of area	
	from Oy	from Ox	about Oy	about Ox
A	x	y	Ax	Ay
$20 \times 30 = 600$	10	15	$600 \times 10 = 6000$	$600 \times 15 = 9000$
$30 \times 60 = 1800$	35	30	$1800 \times 35 = 63\,000$	$1800 \times 30 = 54\,000$
$-\pi \times 10^2 = -310$	35	15	$-310 \times 35 = -10\,900$	$-310 \times 35 = -4700$
$\Sigma A = 2090$			$\Sigma Ax = 58\,100$	$\Sigma Ay = 58\,300$

Hence $$\bar{x} = \frac{\Sigma\,Ax}{\Sigma\,A} = \frac{58\,100}{2090} = 27.8 \text{ mm}$$

and $$\bar{y} = \frac{\Sigma\,Ay}{\Sigma\,A} = \frac{58\,300}{2090} = 27.9 \text{ mm}$$

FINDING CENTROIDS USING INTEGRATION

There are many shapes which cannot be divided up exactly into either rectangles or circles. These more complicated shapes can be split up into elemental strips which approximate to rectangles. Then by finding the dimensions of each strip, etc. the formulae $\bar{x} = \dfrac{\Sigma\,Ax}{\Sigma\,A}$ and $\bar{y} = \dfrac{\Sigma\,Ay}{\Sigma\,A}$ may be used to find the approximate values of $\bar{x}$ and $\bar{y}$.

However if we know, or can find, the equation of y in terms of x, by setting up the area on suitable axes, the numerical summing up may be achieved by integration and exact results obtained.

The following examples will explain how integration is used to find a centroid.

EXAMPLE 2

Find, by integration, the value of $\bar{x}$ (i.e. the distance of the centroid from the left hand edge) for the rectangle shown in Fig. 6.7.

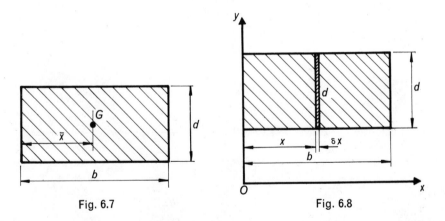

Fig. 6.7   Fig. 6.8

The rectangle must be set up on suitable axes. Fig. 6.8 shows a sketch of the arrangement and includes the usual elementary strip of very small width δx and length d.

Now the area of the rectangle $= \Sigma A$

$\qquad\qquad\quad =$ sum of the elementary strip areas

$$= \sum_{x=0}^{x=b} d.\delta x$$

$$= \int_0^b d.dx$$

$$= d\int_0^b 1.dx$$

$$= d\left[x\right]_0^b = d(b-0) = b.d$$

Also the first moment of the rectangular area about the y-axis

$$= \Sigma A.x$$

$\qquad\qquad\quad =$ sum of the first moment of area of each
of the elementary strips

$\qquad\qquad\quad =$ sum of: (area of strip $\times$ distance of its
centroid from the y-axis)

$$= \sum_{x=0}^{x=b} (d.\delta)x$$

$$= \int_0^b d.x.dx$$

$$= d\int_0^b x.dx$$

$$= d\left[\frac{x^2}{2}\right]_0^b$$

$$= d\left(\frac{b^2}{2} - \frac{0^2}{2}\right)$$

$$= \frac{b^2.d}{2}$$

Hence $\qquad\qquad \bar{x} = \dfrac{\Sigma A.x}{\Sigma A} = \dfrac{b^2 d/2}{bd} = \dfrac{b}{2}$

This is the result we would expect from the symmetry of the figure. The procedure followed in this example is always used when finding positions of centroids by integration.

EXAMPLE 3

Find the position of the centroid of a triangle of height H and base length B.

The triangle must be set up on suitable axes. Only experience enables a good choice. In this case it is convenient to turn the triangle through 90° so that the apex lies on the vertical y-axis and the base is parallel to it. In this instance the position of the x-axis is unimportant.

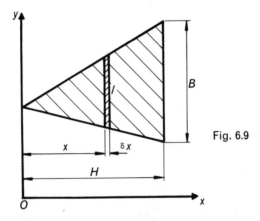

Fig. 6.9

Fig. 6.9 shows a sketch of the arrangement and includes the usual elementary strip area of very small width δx and length l. We shall need a connection between l and x and this may be found using similar triangles

giving $\dfrac{l}{x} = \dfrac{B}{H}$ from which $l = \dfrac{B}{H}.x$.

Now the area of the triangle $= \Sigma A$

$$= \text{sum of the elementary strip areas}$$

$$= \sum_{x=0}^{x=H} l.\delta x$$

$$= \int_0^H l.\mathrm{d}x$$

$$= \int_0^H \frac{B}{H}.x.\mathrm{d}x$$

$$= \frac{B}{H} \int_0^H x.\mathrm{d}x$$

$$= \frac{B}{H}\left[\frac{x^2}{2}\right]_0^H = \frac{B}{H}\left(\frac{H^2}{2} - \frac{0^2}{2}\right) = \frac{1}{2}BH$$

Also the first moment of the triangular area about the y-axis:

$\Sigma\, Ax =$ sum of the first moment of area of each
of the elementary strips

$=$ sum of (area of strip $\times$ distance of its
centroid from the y-axis)

$$= \sum_{x=0}^{x=H} l.\delta x.x$$

$$= \int_0^H l.dx.x$$

$$= \int_0^H \frac{B}{H}.x.dx.x$$

$$= \frac{B}{H} \int_0^H x^2\, dx$$

$$= \frac{B}{H}\left[\frac{x^3}{3}\right]_0^H = \frac{B}{H}\left\{\frac{H^3}{3} - \frac{0^3}{3}\right\} = \frac{1}{3}BH^2$$

Hence $\bar{x}$ $\qquad = \dfrac{\Sigma\, Ax}{\Sigma\, A} = \dfrac{\frac{1}{3}BH^2}{\frac{1}{2}BH} = \dfrac{2}{3}H$

It should be noted that this is independent of the base length B.

You may remember that the position of the centroid of a triangular area is as given in Fig. 6.10. The above calculations verify this.

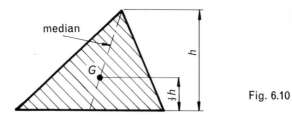

Fig. 6.10

EXAMPLE 4

Find the position of the centroid of the sector of a circle of radius R and angle 2α.

The sector area is shown in Fig. 6.11 located on suitable axes Ox and Oy. By symmetry the centroid G of the sector area lies on the centre-line OM, and its position along the line may be defined if we find the distance $\bar{x}$.

It is not convenient to divide the area into vertical (or even horizontal) elementary strip areas. Instead we consider elementary sector areas of small angle $\delta\theta$ (Fig. 6.12). Each of these elementary areas may be considered to approximate to a triangle having base length $R \times \delta\theta$ (since

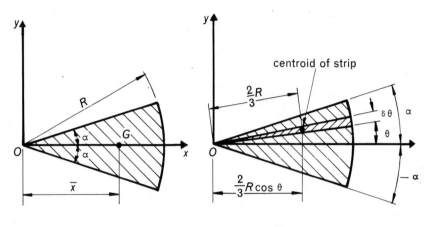

Fig. 6.11   Fig. 6.12

length of arc = angle × radius, where the angle is in radians), and height R.

We must first find the area of the sector by summing the areas of the elementary sector areas.

Hence   area of sector = $\Sigma\, A$

$\qquad\qquad$ = sum of elementary sector areas

$$= \sum_{\theta=-\alpha}^{\theta=\alpha} \frac{1}{2}(R.\delta\theta)R \qquad \text{since the area of a triangle is } \frac{1}{2}(\text{base})(\text{height})$$

$$= \int_{-\alpha}^{\alpha} \frac{1}{2}(R.\mathrm{d}\theta)R$$

$$= \tfrac{1}{2}R^2 \int_{-\alpha}^{\alpha} \mathrm{d}\theta$$

$$= \tfrac{1}{2}R^2 \Big[\,\theta\,\Big]_{-\alpha}^{\alpha} = \frac{1}{2}R^2\{\alpha-(-\alpha)\}$$

$$= R^2\alpha$$

Now the first moment of the given sector area about the y-axis:

$\Sigma\, Ax$ = sum of the first moment of area of each of the elementary sector areas

$\qquad\qquad$ = sum of (elementary area × distance of its centroid from the y-axis)

$$= \sum_{\theta=-\alpha}^{\theta=\alpha} (\tfrac{1}{2}R^2\delta\theta)(\tfrac{2}{3}R.\cos\theta)$$

$$= \int_{-\alpha}^{\alpha} \tfrac{1}{2}R^2 . d\theta . \tfrac{2}{3} R . \cos \theta$$

$$= \tfrac{1}{3}R^3 \int_{-\alpha}^{\alpha} \cos \theta \; d\theta$$

$$= \tfrac{1}{3}R^3 \Big[\sin \theta \Big]_{-\alpha}^{\alpha}$$

$$= \tfrac{1}{3}R^3 \{ \sin \alpha - \sin(-\alpha) \}$$

$$= \tfrac{1}{3}R^3 \{ \sin \alpha + \sin \alpha \}$$

$$= \tfrac{2}{3}R^3 \sin \alpha .$$

Now

$$\bar{x} = \frac{\Sigma \, Ax}{\Sigma \, A}$$

$$= \frac{\tfrac{2}{3}R^3 . \sin \alpha}{R^2 \alpha}$$

$$= \frac{2}{3} R \frac{\sin \alpha}{\alpha}$$

It should be remembered that the angle α must be in radians.

EXAMPLE 5

Using the result obtained for the position of the centroid of the sector of a circle find:

(a)   the position of the centroid of a quadrant of a circle of radius R.

(b)   the position of the centroid of a semicircular area of radius R.

(a)   If we set the quadrant area on suitable axes as shown in Fig. 6.13, then we may use the formula for the sector of a circle giving:

$$OG = \tfrac{2}{3}R \frac{\sin}{\alpha} \alpha \quad \text{where} \quad \alpha = \frac{\pi}{4} \text{ rad}$$

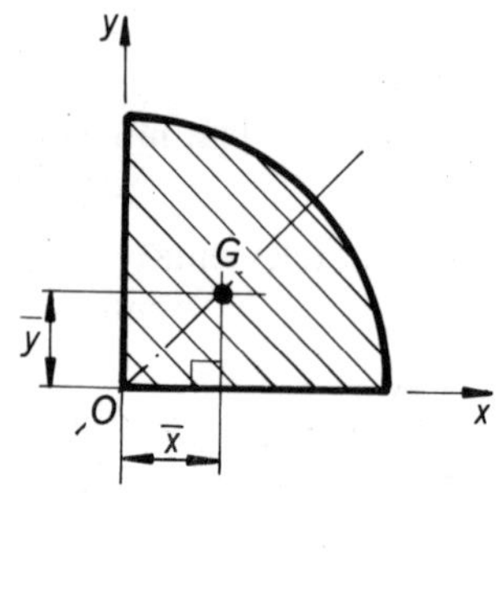

Fig. 6.13

From right angled triangle OGM we have:

$$\bar{x} = OM$$

$$= OG\left(\cos\frac{\pi}{4}\right)$$

$$= \tfrac{2}{3}R\,\frac{\sin\frac{\pi}{4}}{\frac{\pi}{4}}\left(\cos\frac{\pi}{4}\right)$$

$$= \frac{2R(0.7071)(0.7071)4}{3\pi}$$

$$= \frac{4R}{3\pi}$$

$$= 0.424R$$

Also by symmetry $\bar{y}$ will have the same value as $\bar{x}$.

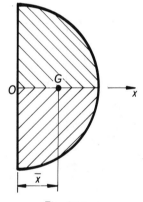

Fig. 6.14

(b) The semi-circular area shown in Fig. 6.14 can be considered to be two quadrants as indicated. It follows, therefore, that $\bar{x}$ for a semi-circular area is the same as that for a quadrant, i.e. $0.424R$.

By symmetry the centroid also lies on the horizontal centre-line Ox.

Exercise 13

In Questions 1–4 find the positions of the centroids of the cross-sectional areas shown, giving the distances in each case from the left hand edge (or extreme point) and above the bottom edge (or extreme point).

1)

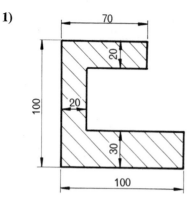

2)

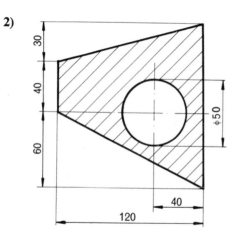

3)

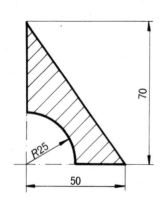

4)

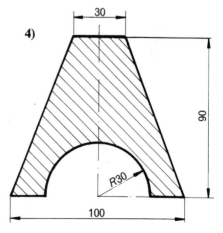

In Questions 5–8 find $\bar{x}$ and $\bar{y}$ as indicated on the diagrams.

5)

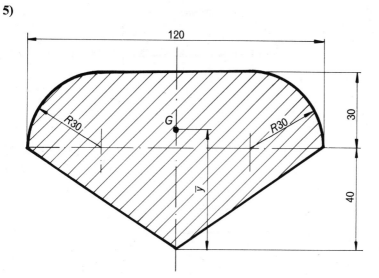

6)

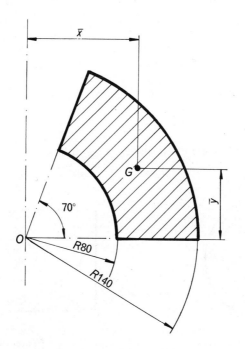

7)

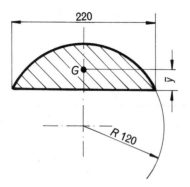

8)

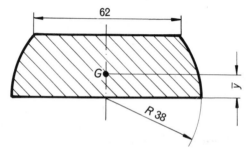

SUMMARY

Centroids of the more common shapes are as follows:

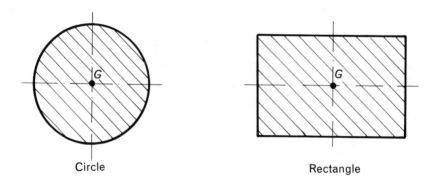

Circle   Rectangle

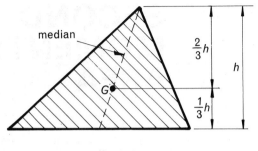

median

$\frac{2}{3}h$

h

$\frac{1}{3}h$

G

Triangle

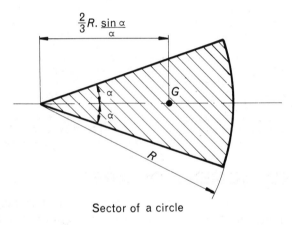

$\frac{2}{3}R.\frac{\sin\alpha}{\alpha}$

α

α

G

R

Sector of a circle

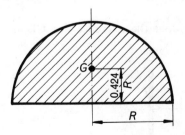

G

0.424 R

R

Semi-circle

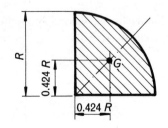

R

0.424 R

G

0.424 R

Quadrant of a circle

SECOND MOMENTS OF AREA

On reaching the end of this chapter you should be able to :-

1. Sketch the given area including a typical incremental area parallel to a specified axis in the plane of the area.
2. Define the second moment of area.
3. Determine the second moment of the increment in 1 about the specified axis.
4. Determine the second moment of area by summing the second moment of the incremental area between given limits by definite intergation.
5. Determine the second moment of a rectangle about one edge, a triangle about one edge, a circle and semi-circle about a diameter, and a circle about its polar axis.
6. State the parallel axis theorem.
7. Determine the second moment of area about an axis through the centroid of the area parallel to

the axis about which the second moment is known using 6.
8. Determine the second moment of area about an axis parallel to the axis about which the second moment is known using 6 and 7.
9. Calculate the second moments of the component areas of a composite area about a common axis using 8.
10. Calculate the second moment of a composite area by summing the second moments in 9.
11. State the perpendicular axis theorem.
12. Calculate the second moments of area of a given area about two axes at right angles in the plane of the given area.
13. Calculate the second moment of area about a perpendicular axis through the intersection of axes in 12 using 11.

SECOND MOMENT OF AREA

The second moment of area is a property of an area used in many engineering calculations. One example is in finding stresses, due to bending, which involves the use of the second moment of the cross-sectional area of the beam.

Fig. 7.1 shows a thin strip, of area A and of very small width, which is distant x from the reference line CD.

Now the first moment of the area A about the reference line CD is given by $A.x$.

Similarly the second moment of area A about the reference line CD is given by $A.x^2$.

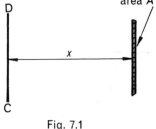

Fig. 7.1

The symbol for the second moment of area is I and is always stated with reference to an axis or datum line. Thus for the strip area shown

$$I_{CD} = A.x^2$$

The expression $A.x^2$ is only true if the area is a very thin strip parallel to the reference line. It follows that if we require the second moment of any other shaped area it is necessary to divide the area into a number of elementary strips all parallel to the reference line.

Thus in Fig. 7.2 for the irregular area $I_{CD} = \Sigma A.x^2$.

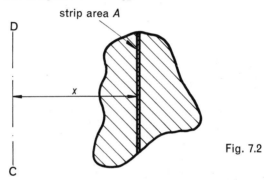

strip area A

Fig. 7.2

The summation of the $A.x^2$ for each strip may be carried out by graphical and numerical methods but, as in the case of finding first moments of area, this summation may often be achieved by integration as the following example will illustrate:

EXAMPLE 1

Find the second moment of area of the rectangle shown in Fig. 7.3 about its base edge.

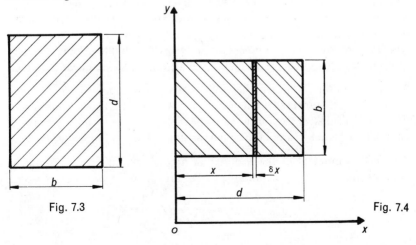

Fig. 7.3

Fig. 7.4

The rectangular area must be set up on suitable axes. In this case it is convenient to turn the rectangle through 90° and arrange the base edge to lie on the y-axis as shown in Fig. 7.4. In this instance the position of the x-axis is unimportant.

The diagram shows a typical elementary strip area parallel to the reference axis and whose area is $b.\delta x$.

Now the second moment of the rectangular area:

$$\text{about the } y\text{-axis} = \Sigma\, A.x^2$$

$$= \sum_{x=0}^{x=d} b.\delta x.x^2$$

$$= \int_0^d b.\mathrm{d}x.x^2$$

$$= b \int_0^d x^2\, \mathrm{d}x$$

$$= b\left[\frac{x^3}{3}\right]_0^d = b\left(\frac{d^3}{3} - \frac{0^3}{3}\right) = \frac{bd^3}{3}$$

UNITS OF SECOND MOMENTS OF AREA

The second moment of area is $\Sigma\, A.x^2$. If all lengths are in metres, then the result will be $m^2 \times m^2 = m^4$.

For example in the formula $\dfrac{bd^3}{3}$ the units are $m \times m^3 = m^4$.

In typical engineering problems it is unlikely that dimensions of a cross-sectional area will be in metres. It is more probable that the units will be millimetres, and if these are used the second moment of area will be in mm^4 which will result in large numbers. In practice it is usual to calculate second moments of area in cm^4. This makes the arithmetic as simple as possible.

This is one of the few times when units not recommended by the Système International (SI) are used in engineering calculations. In British industry producers of rolled steel sections used for beams and columns, etc. agreed with continental manufacturers to use cm^4 as units of second moments of area.

The second moment of the rectangular area shown in Fig. 7.5 about its base, or more correctly about an axis BB on which the base lies, is,

$$I_{BB} = \frac{bd^3}{3} = \frac{3 \times 2^3}{3} = 8 \text{ cm}^4$$

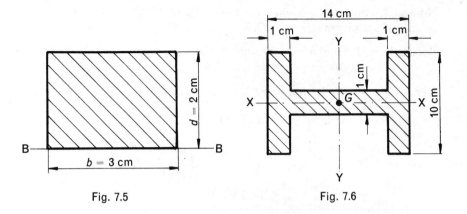

Fig. 7.5   Fig. 7.6

CONVENTIONAL NOTATION I_{XX} AND I_{YY}

XX and YY are known conventionally as the horizontal and vertical axes which pass through the centroid of a cross-sectional area.

Hence: I_{XX} is the second moment of a sectional area about the horizontal axis which passes through the centroid,

and: I_{YY} is the second moment of a sectional area about the vertical axis which passes through the centroid.

EXAMPLE 2

Find I_{XX} and I_{YY} of the cross-sectional area shown in Fig. 7.6.

In this case by symmetry the centroid G will be at the centre of the section.

To find I_{XX}

We shall consider the area to be made up by six rectangles, as shown in Fig. 7.7, each of which has its 'base' on XX.

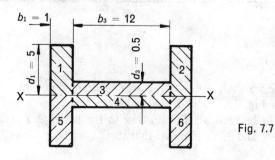

Fig. 7.7

Now I_{XX} for the cross-sectional area will be the sum of the second moments of area of each rectangle about XX.

Rectangles 1, 2, 5 and 6 have the same dimensions and hence the same I_{XX}. Similarly rectangles 3 and 4 have the same I_{XX}.

Hence I_{XX} for whole area $= 4(I_{XX}$ for area 1$)+2(I_{XX}$ for area 3$)$

$$= 4\left(\frac{b_1 d_1^3}{3}\right)+2\left(\frac{b_3 d_3^3}{3}\right)$$

$$= 4\left(\frac{1\times 5^3}{3}\right)+2\left(\frac{12\times 0.5^3}{3}\right)$$

$$= 166.7+1.0$$

$$= 167.7 \text{ cm}^4$$

To find I_{YY}

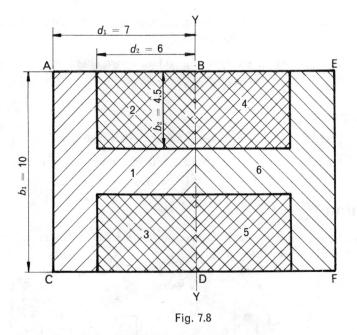

Fig. 7.8

As before we shall consider the area to be made up by six rectangles, as shown in Fig. 7.8, each of which has its 'base' on YY. In this case the 'bases' will be vertical. The portion to the left of YY is area 1 (ABDC) less areas 2 and 3. This means that the I_{YY} of both areas 2 and 3 must be taken as negative when finding I_{YY} for the whole section. We treat the portion of the cross-section to the right of YY in a similar manner.

Rectangles 1 and 6 have the same dimensions and hence the same I_{YY}. Similary rectangles 2, 3, 4 and 5 have the same I_{YY}.

Hence I_{YY} for whole area = $2(I_{YY}$ for area 1)$-4(I_{YY}$ for area 2)

$$= 2\left(\frac{b_1 d_1{}^3}{3}\right) - 4\left(\frac{b_2 d_2{}^3}{3}\right)$$

$$= 2\left(\frac{10 \times 7^3}{3}\right) - 4\left(\frac{4.5 \times 6^3}{3}\right)$$

$$= 2287 - 1296$$

$$= 991 \ \text{cm}^4$$

EXAMPLE 3

Find I_{XX} and I_{YY} for the cross-sectional area shown in Fig. 7.9.

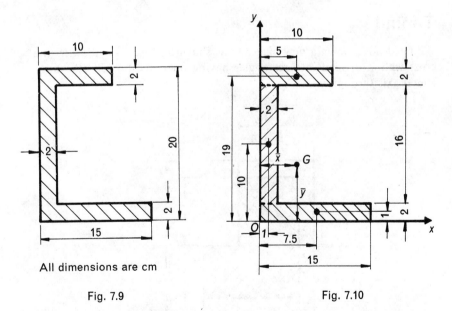

All dimensions are cm

Fig. 7.9   Fig. 7.10

First it is necessary to find the position of the centroid G. We will divide the cross-section into three rectangular areas as shown in Fig. 7.10 and then find the position of G relative to the chosen reference axes Ox and Oy.

Area	Distance to centroid		First moment of area	
	from Oy	from Ox	about Oy	about Ox
A	x	y	Ax	Ay
$10 \times 2 = 20$	5	19	$20 \times 5 = 100$	$20 \times 19 = 380$
$2 \times 16 = 32$	1	10	$32 \times 1 = 32$	$32 \times 10 = 320$
$15 \times 2 = 30$	7.5	1	$30 \times 7.5 = 225$	$30 \times 1 = 30$
$\Sigma A = 82$			$\Sigma Ax = 357$	$\Sigma Ay = 730$

Hence
$$\bar{x} = \frac{\Sigma Ax}{\Sigma A} = \frac{357}{82} = 4.35 \text{ cm}$$

and
$$\bar{y} = \frac{\Sigma Ay}{\Sigma A} = \frac{730}{82} = 8.90 \text{ cm}$$

To find I_{xx}

We must now sketch a new diagram, as shown in Fig. 7.11, showing the cross-section divided into rectangles having one edge on **XX** (the horizontal axis through G).

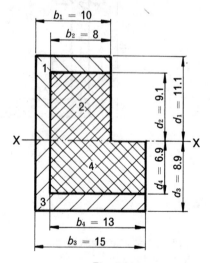

Fig. 7.11

Hence,

I_{XX} for whole area $= I_{XX}$ for area $1-I_{XX}$ for area $2+I_{XX}$ for area 3

$$-I_{XX} \text{ for area } 4$$

$$= \frac{b_1 d_1^3}{3} - \frac{b_2 d_2^3}{3} + \frac{b_3 d_3^3}{3} - \frac{b_4 d_4^3}{3}$$

$$= \frac{10(11.1)^3}{3} - \frac{8(9.1)^3}{3} + \frac{15(8.9)^3}{3} - \frac{13(6.9)^3}{3}$$

$$= 4560 - 2010 + 3520 - 1420$$

$$= 4650 \text{ cm}^4$$

To find I_{YY}

Again a new sketch is required, as shown in Fig. 7.12, showing the cross-section divided into rectangles having one edge on YY (the vertical axis through G).

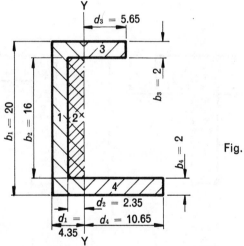

Fig. 7.12

Hence,

I_{YY} for whole area $= I_{YY}$ for area $1-I_{YY}$ for area $2+I_{XX}$ for area 3

$$+I_{YY} \text{ for area } 4$$

$$= \frac{b_1 d_1^3}{3} - \frac{b_2 d_2^3}{3} + \frac{b_3 d_3^3}{3} + \frac{b_4 d_4^3}{3}$$

$$= \frac{20(4.35)^3}{3} - \frac{16(2.35)^3}{3} + \frac{2(5.65)^3}{3} + \frac{2(10.65)^3}{3}$$

$$= 549 - 69 + 120 + 805$$

$$= 1405 \text{ cm}^4$$

Exercise 14

Find I_{XX} and I_{YY} for the cross-sectional areas given in the following examples. All dimensions are cm.

1)

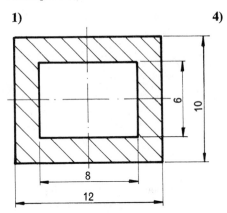

4)

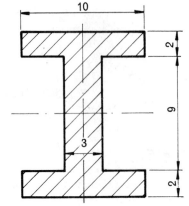

2)

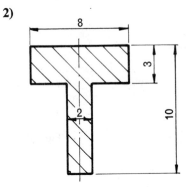

5)

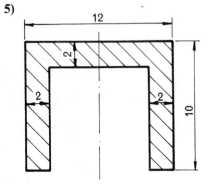

3)

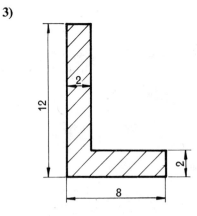

6)

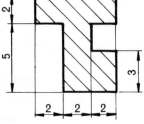

EXAMPLE 4

Find the second moment of a triangular area, of height H and base B, about a line through its apex and parallel to the base.

The arrangement of the axes etc. is identical to that used in Example 3, p. 75, and is as shown in Fig. 7.13.

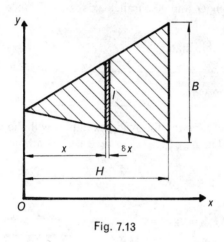

Fig. 7.13

Now the second moment of the triangular area about the y-axis:

$$= \Sigma\, A.x^2$$

$$= \sum_{x=0}^{x=H} l.\delta x.x^2$$

$$= \int_0^H l.x^2\, \mathrm{d}x$$

and using similar triangles as before we have $l = \dfrac{B}{H}.x$

$$\therefore \qquad I_{Oy} = \int_0^H \frac{B}{H}.x.x^2\, \mathrm{d}x$$

$$= \frac{B}{H} \int_0^H x^3\, \mathrm{d}x$$

$$= \frac{B}{H}\left[\frac{x^4}{4}\right]_0^H = \frac{B}{H}\left\{\frac{H^4}{4} - \frac{0^4}{4}\right\} = \frac{B}{H}.\frac{H^4}{4} = \frac{BH^3}{4}$$

THE PARALLEL AXIS THEOREM

A second moment of area must always be stated together with the axis about which it has been calculated. We state by saying, for example, that I_{AB} is the second moment of area about the axis AB.

Fig. 7.14 shows a plane area A whose centroid is G. Also shown is an axis which passes through G and a parallel axis, the distance between the axes being h.

The parallel axis theorem states:

$$I_{\text{parallel axis}} = I_{\text{axis throught } G} + A.h^2$$

This is extremely useful as the following examples will show. It is worth while remembering that $I_{\text{axis through } G}$ is the least second moment of area.

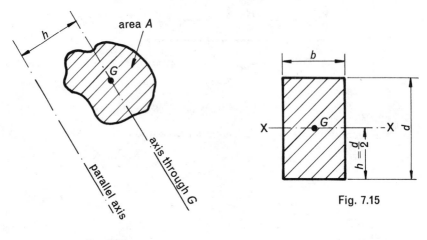

area A

Fig. 7.14

Fig. 7.15

EXAMPLE 5

Find the second moment of area of the rectangle shown in Fig. 7.15 about the axis XX which passes through the centroid.

We know from Example 1 that the second moment of area of the rectangle about its base is $\dfrac{b.d^3}{3}$.

Now using the parallel axis theorem we have

$$I_{\text{base}} = I_{XX} + A.h^2$$

∴ rearranging

$$I_{XX} = I_{base} - A.h^2$$

$$= \frac{bd^3}{3} - bd\left(\frac{d}{2}\right)^2$$

$$= bd^3\left(\frac{1}{3} - \frac{1}{4}\right)$$

$$= \frac{bd^3}{12}$$

For many of the shapes used in engineering a tabular method using the theorem of parallel axes is often used as an alternative to the method used in Examples 2 and 3.

EXAMPLE 6

Find I_{XX} for the cross-section shown in Fig. 7.16.

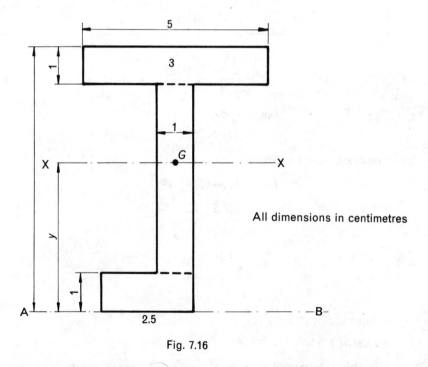

All dimensions in centimetres

Fig. 7.16

The first step is to choose a suitable axis, such as AB in Fig. 7.16, and calculate I_{AB}. To do this the cross-section is divided into the three rect-angles 1, 2 and 3. Note that for each rectangle $I_{AB} = Ay^2 + I_G$

where A = area of the rectangle

y = distance of centroid of the rectangle from AB

I_G = second moment of area of the rectangle about
the centroid of the rectangle $= \dfrac{bd^3}{12}$

Item	A	y	Ay	Ay^2	$I_G = \dfrac{bd^3}{12}$
1	2.5	0.5	1.25	0.6	$\dfrac{2.5 \times 1^3}{12} = 0.2$
2	5.0	3.5	17.50	61.3	$\dfrac{1 \times 5^3}{12} = 10.4$
3	5.0	6.5	32.50	211.3	$\dfrac{5 \times 1^3}{12} = 0.4$
	12.5		51.25	273.2	11.0

$$\bar{y} = \frac{\Sigma\, Ay}{\Sigma\, A} = \frac{51.25}{12.5} = 4.10 \text{ cm}$$

∴ total $I_{AB} = \Sigma\, Ay^2 + \Sigma\, I_G$

$$= 273.2 + 11.0 = 284.2 \text{ cm}^4$$

and again using the parallel axis we have:

$$I_{XX} = I_{AB} - (\Sigma\, A)x\bar{y}^2$$

$$= 284.2 - 12.5 \times (4.10)^2$$

$$= 284.2 - 210.1$$

$$= 74.1 \text{ cm}^4$$

EXAMPLE 7

Given that the second moment of a triangular area, of height H and base B, about a line through its apex and parallel to its base is $\dfrac{B.H^3}{4}$ find the second moment of the triangular area about its base.

Fig. 7.17 shows a sketch of the triangular area with axes suitably labelled. We are given I_{MM} and we require I_{NN}.

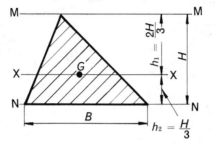

Fig. 7.17

The parallel axis theorem requires that one of the axes used must pass through the centroid. It is necessary, therefore, to use the parallel axis theorem twice — once to find I_{XX} knowing I_{MM} and again to find I_{NN} using I_{XX}.

Now the parallel axis theorem gives $I_{MM} = I_{XX} + A.h^2$

and rearranging $I_{XX} = I_{MM} - A.h^2{}_1$

$\therefore$ $$I_{XX} = \frac{B.H^3}{4} - \frac{B.H}{2}\left(\frac{2H}{3}\right)^2$$

$$= \frac{B.H^3}{36}$$

Also the parallel axis theorem gives $I_{NN} = I_{XX} + A.h_2{}^2$

$$= \frac{B.H^3}{36} + \frac{B.H}{2}\left(\frac{H}{3}\right)^2$$

$$= \frac{B.H^3}{12}$$

Hence the second moment of a triangular area about its base is $\dfrac{B.H^3}{12}$.

EXAMPLE 8

The second moment of a rectangular area, of base length b and depth d, about an axis through its centroid and parallel to the base is $\dfrac{b.d^3}{12}$. The second moment of a triangular area, of height H and base length B, about an axis along its base is $\dfrac{B.H^3}{12}$.

Using the above data and also the parallel axis theorem for the area shown in Fig. 7.18, find:

(a) I_{XX} (b) I_{YY} (c) I_{CD}

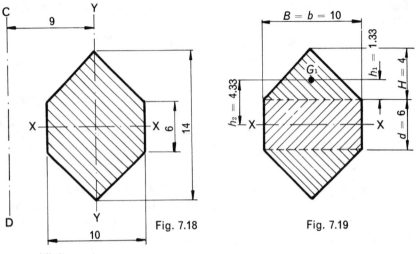

Fig. 7.18

Fig. 7.19

All dimensions are cm

(a) To find I_{XX}

Fig. 7.19 shows the given area divided into a rectangle and two triangles. Since the triangles are of identical dimensions and the same disposition relative to the axis XX then their second moments of area about XX will be the same. The parallel axis theorem will be used to find I_{XX} of the triangular areas about XX knowing their I_{base}. This will have to be done in two stages:

(i)   To find I of the triangle about a line parallel to the base of the triangle passing through G_T (G_T being the centroid of the triangle).

(ii)   To find I_{XX} for the triangle.

For the triangular area the parallel axis theorem gives

$$I_{base} = I_{G_T} + (\text{area of triangle}) \times h_1{}^2$$

∴ rearranging

$$I_{G_T} = I_{base} - (\text{area of triangle}) \times h_1{}^2$$

$$= \frac{B.H^3}{12} - \frac{1}{2}.B.H \times h_1{}^2$$

$$= \frac{10 \times 4^3}{12} - \frac{1}{2} \times 10 \times 4 \times 1.33^2$$

$$= 53.3 - 35.4$$

$$= 17.9 \text{ cm}^4$$

For the same triangular area the parallel axis theorem gives

$$I_{XX} = I_{G_T} + (\text{area of triangle}) \times h_2{}^2$$

$$= 17.9 + \left(\frac{1}{2} \times 10 \times 4\right) \times 4.33^2$$

$$= 17.9 + 375$$

$$= 393 \text{ cm}^4$$

Now for the given area

$$I_{XX} \text{ of given area} = I_{XX} \text{ of rectangle} + 2(I_{XX} \text{ of triangle})$$

$$= \frac{b.d^3}{12} + 2(393)$$

$$= \frac{10 \times 6^3}{12} + 786$$

$$= 180 + 786$$

$$= 966 \text{ cm}^4$$

(b) To find I_{YY}

Fig. 7.20 shows the given area divided into a rectangle and four triangles of equal dimensions.

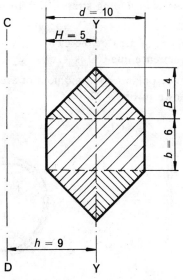

Fig. 7.20

Hence I_{YY} of given area $= I_{YY}$ of rectangle$+4(I_{YY}$ of triangle)

$$= \frac{b.d^3}{12}+4\left(\frac{B.H^3}{12}\right)$$

$$= \frac{6\times 10^3}{12}+4\left(\frac{4\times 5^3}{12}\right)$$

$$= 500+167$$

$$= 667 \text{ cm}^4$$

(c) I_{CD} may be found using the parallel axis theorem for the given area.

Hence I_{CD} of given area $= I_{YY}$ of given area$+$(given area)$\times h^2$

$$= 667+\left(6\times 10+2\times\frac{1}{2}\times 10\times 4\right)\times 9^2$$

$$= 667+8100$$

$$= 8767 \text{ cm}^4$$

POLAR SECOND MOMENT OF AREA

The polar second moment of area, denoted by the symbol J, is the second moment of a circular area about the polar axis (that is the axis passing through the centre of the area and perpendicular to the plane of the area).

One example of its use is in finding stresses due to torsion in a circular shaft.

Now the second moment of area of the elementary strip area, shown in Fig. 7.21, about the reference line CD is $A.x^2$.

Similarly we say that the polar second moment of area of the elementary circular strip area, shown in Fig. 7.22, is $A.r^2$.

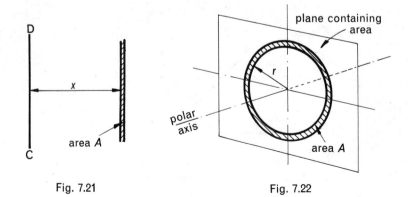

Fig. 7.21   Fig. 7.22

EXAMPLE 9

Find the polar second moment of a circular area of diameter D.

Fig. 7.23 shows the given area and an elementary circular strip. We shall sum the $A.x^2$ for each strip area to find the polar second moment of area for the whole area.

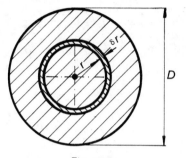

Fig. 7.23

Hence J for the circular area $= \Sigma A.r^2$

Now the approximate area A of the elementary circular strip

$$= \text{(circumference of strip)} \times \text{(width of strip)}$$

$$= 2\pi r \times \delta r$$

$\therefore$ J for the circular area $= \displaystyle\sum_{r=0}^{r=D/2} 2\pi r.\delta r.r^2$

$$= \int_0^{D/2} 2\pi r.dr.r^2$$

$$= 2\pi \int_0^{D/2} r^3 \, dr$$

$$= 2\pi \left[\frac{r^4}{4}\right]_0^{D/2}$$

$$= 2\pi \left\{\frac{(D/2)^4}{4} - \frac{0^4}{4}\right\}$$

$$= \frac{\pi D^4}{32}$$

EXAMPLE 10

Find the polar second moment of area of the cross-section of a tube 6 cm outside diameter and 4 cm inside diameter.

We shall consider J of the cross-sectional area shown in Fig. 7.24 to be J of a 6 cm diameter circular area less J of a 4 cm diameter circular area.

$\therefore$ $\dfrac{\pi(d^4 - d^4)}{64}$ required $J = \dfrac{\pi D^4}{32} - \dfrac{\pi d^4}{32}$

$\dfrac{\pi(1296 - 256)}{64}$ $= \dfrac{\pi 6^4}{32} - \dfrac{\pi 4^4}{32}$

$I_{xx} \quad = 51 + 51$ $= 127 - 25$

J $= 102$ cm^4

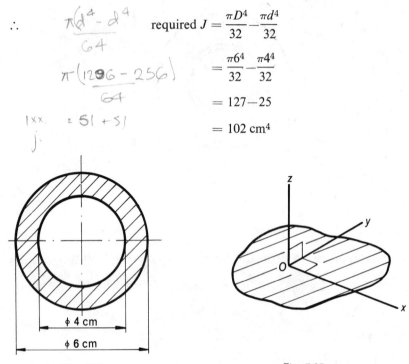

Fig. 7.24   Fig. 7.25

THE PERPENDICULAR AXIS THEOREM

Fig. 7.25 shows a plane area on which the axes Ox and Oy are drawn, the axes being at right angles.

Oz is an axis perpendicular to the plane area. Ox, Oy and Oz are said to be mutually perpendicular.

The perpendicular axis theorem states:

$$I_{Oz} = I_{Ox} + I_{Oy}$$

EXAMPLE 11

Find I_{Oz} for the rectangle shown in Fig. 7.26.

We know that for a rectangle

$$I_{\text{base}} = \frac{b.d^3}{3}$$

$$\therefore \qquad I_{OA} = \frac{12 \times 8^3}{3} = 2048 \text{ cm}^4$$

and $\qquad I_{OB} = \frac{8 \times 12^3}{3} = 4608 \text{ cm}^4$

Now the perpendicular axis theorem states:

$$I_{Oz} = I_{OA} + I_{OB}$$
$$= 2048 + 4608$$
$$= 6656 \text{ cm}^4$$

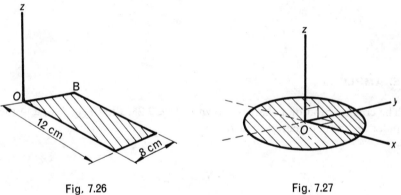

Fig. 7.26   Fig. 7.27

EXAMPLE 12

Using the fact that J for a circular area is $\dfrac{\pi d^4}{32}$ and the perpendicular axis theorem find I about a diameter.

Now for a circular area I_{Oz} is called J (Fig. 7.27), and both I_{Ox} and I_{Oy} are equal to I_{diameter}.

The perpendicular axis theorem states:

$$I_{Oz} = I_{Ox} + I_{Oy}$$

$$\therefore \qquad J = I_{\text{diameter}} + I_{\text{diameter}}$$

hence
$$I_{\text{diameter}} = \frac{1}{2} \cdot J$$

$$= \frac{1}{2} \cdot \frac{\pi d^4}{32}$$

$$= \frac{\pi d^4}{64}$$

Hence I about a diameter of a circular area is $\dfrac{\pi d^4}{64}$.

A semi-circular area has I about the boundary diameter equal to one half of the result obtained for a full circular area about a diameter, hence for a semi-circular area:

$$I_{\text{diameter}} = \frac{\pi d^4}{128}$$

EXAMPLE 13

The cross-section of a shaft is shown in Fig. 7.28. Find I_{XX}, I_{YY}, and the polar second moment of area J.

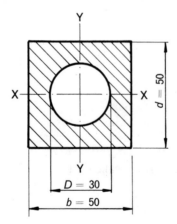

Fig. 7.28

To find I_{XX}

$$I_{XX} \text{ of cross-section} = I_{XX} \text{ of square} - I_{XX} \text{ of circle}$$

$$= \frac{b.d^3}{12} - \frac{\pi D^4}{64}$$

$$= \frac{50 \times 50^3}{12} - \frac{\pi 30^4}{64}$$

$$= 520\,800 - 39\,800$$

$$= 481\,000 \text{ mm}^4$$

$$= \frac{481\,000}{10^4} \text{ cm}^4 = 48.1 \text{ cm}^4$$

To find I_{YY}

We may see from the symmetry of the figure that $I_{YY} = I_{XX}$

Hence $I_{YY} = 48.1 \text{ cm}^4$.

To find J

Using the perpendicular axis theorem we have:

$$J = I_{XX} + I_{YY}$$

$$= 48.1 + 48.1$$

$$= 96.2 \text{ cm}^4$$

RADIUS OF GYRATION

The formula for finding centroids is:

$$(\Sigma A).\bar{x} = \Sigma A.x$$

Similarly for second moments of area we have:

$$I = (\Sigma A).k^2 = \Sigma A.x^2$$

where k is called the 'radius of gyration' of the area.

In this chapter we have concentrated on using I found from $\Sigma A.x^2$. However we are given sometimes the area, ΣA, of a cross-section together with the radius of gyration, k, and we then find the second moment of area, I, by using $I = (\Sigma A).k^2$.

EXAMPLE 14

Find I_{XX} for a cross-sectional area given that the area of the section is 85 cm and $k_{XX} = 3.7$ cm.

We have $\qquad\qquad I = (\text{area of section}).k^2$

$\therefore \qquad\qquad I_{XX} = 85(3.7)^2$

$\qquad\qquad\qquad = 1164 \text{ cm}^4$

Exercise 15

Find I_{XX} and I_{YY} for the cross-sectional areas in Questions 1, 2, 3 and 4. All dimensions are cm.

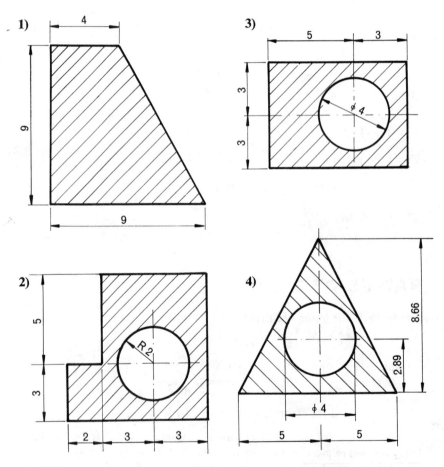

5) Using a tabular method find I_{XX} for the sections shown in Questions 2, 3, 4 and 5 of Exercise 14.

6) For the sectional area shown in Fig. 7.29 find:

(a) *I* about a diameter,

(b) *J* about the polar axis,

(c) I_{AA}.

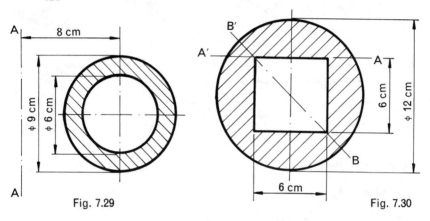

Fig. 7.29   Fig. 7.30

7) Starting with the knowledge that the second moment of area of a rectangle about an edge is $b.d^3/3$ and that the second moment of area of a circle about a diameter is $\pi d^4/64$, and also using the perpendicular and parallel axis theorems, find the second moment of the area shown in Fig. 7.30 about the:

(a) axis A'A,

(b) axis through *G*, perpendicular to the plane of the area,

(c) axis B'B.

SUMMARY

Second moments of the more common areas:

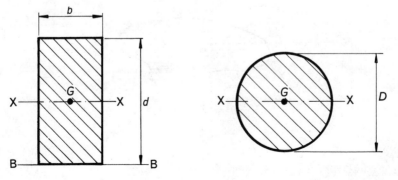

$$I_{XX} = \frac{b.d^3}{12} \quad I_{BB} = \frac{b.d^3}{3} \qquad I_{XX} = \frac{D^4}{64} \quad J = \frac{D^4}{32}$$

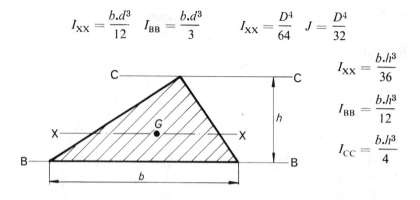

$$I_{XX} = \frac{b.h^3}{36}$$

$$I_{BB} = \frac{b.h^3}{12}$$

$$I_{CC} = \frac{b.h^3}{4}$$

Parallel Axis theorem

$$I_{BB} = I_{XX} + A.h^2$$

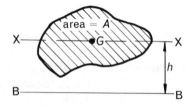

Perpendicular axis theorem

$$I_{ZZ} = I_{XX} + I_{YY}$$

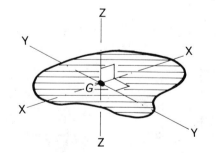

DIFFERENTIAL EQUATIONS

On reaching the end of this chapter you should be able to :-

1. *Determine and sketch a family of curves given their derivative, for a simple function.*
2. *Determine a particular curve of the family by specifing a point on it.*
3. *Define a boundary condition.*
4. *Solve differential equations of the type $\dfrac{dy}{dx} = f(x)$ given a boundary condition.*
5. *Differentiate* $y = Ae^{kx}$.

6. *Verify that* $y = Ae^{kx}$ *satisfies* $\dfrac{dy}{dx} = ky$ *by substitution.*
7. *Derive equations of the form* $\dfrac{dy}{dx} = ky$ *from problems arising in technology.*
8. *Solve the derived equations in 7 using 5 and 6 and a boundary condition.*

FAMILIES OF CURVES

Suppose we know that $\quad \dfrac{dy}{dx} = 3$

This may be rewritten as $\quad y = \displaystyle\int 3\,dx$

from which $\quad y = 3x + C$

The constant of integration C represents any number.

Now suppose we give C different values and plot the graph of each equation (Fig. 8.1).

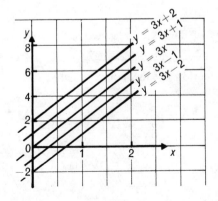

Fig. 8.1

109

When $C = -2$ then $y = 3x - 2$

and when $C = -1$ then $y = 3x - 1$

and when $C = 0$ then $y = 3x$

and when $C = +1$ then $y = 3x + 1$

and when $C = +2$ then $y = 3x + 2$

It follows that the equation $y = 3x + C$ represents a set of graphs called a 'family'. They all have one thing in common, that is $\dfrac{dy}{dx} = 3$.

If we specify a particular point we may find the equation of the graph which passes through that point.

EXAMPLE 1

Find the equation of the graph which is one of the family represented by the equation $y = 3x + C$ and which passes through the point $(1, 7)$.

If a point lies on a graph its co-ordinates satisfy the equation of the graph.

Hence since the point $(1, 7)$ lies on the graph $y = 3x + C$

then $7 = 3(1) + C$

from which $C = 4$

and therefore the required equation is $y = 3x + 4$.

EXAMPLE 2

Sketch the family of curves represented by the equation $\dfrac{dy}{dx} = 2x + 1$.

Find also the equation of the curve which passes through the point $(3, 18)$.

We have, $\dfrac{dy}{dx} = 2x + 1$

or $y = \int (2x + 1)\, dx$

from which $y = x^2 + x + C$

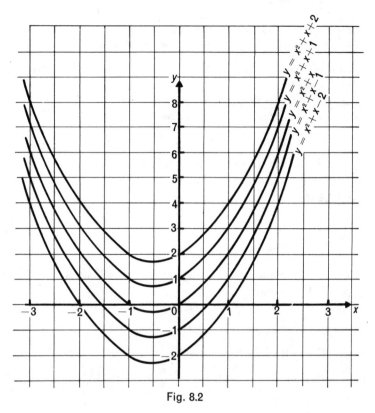

Fig. 8.2

This is the general solution of the given equation and represents the family of curves of which five typical ones are shown in Fig. 8.2.

To find the equation of the curve passing through the point (3, 18) we must substitute these values of x and y in the general solution.

Hence $$18 = 3^2 + 3 + C$$

$\therefore$ $$C = 6$$

and therefore the required equation is $y = x^2 + x + 6$.

DIFFERENTIAL EQUATIONS

Equations which contain a differential coefficient such as:

$$\frac{dy}{dx} = 3 \quad \text{or} \quad \frac{dy}{dx} = 2x + 1$$

are called **differential equations**.

The expressions obtained by integration, for y in terms of x, and which include unknown constants, are called **general solutions**.

If a value of y corresponding to a value of x is known then this is called a **boundary condition.** Values given by a boundary condition may be substituted into a general solution and a numerical value obtained for the constant. The resulting equations, such as:

$$y = 3x+4 \quad \text{and} \quad y = x^2+x+6$$

are called **particular solutions.**

EXAMPLE 3

(a) Find the general solution of the differential equation $\dfrac{dy}{dx} = 5x^3+2x-7.$

(b) From the boundary condition $y = 2$ when $x = 1$, find the particular solution.

(a) We have, $$\frac{dy}{dx} = 5x^3+2x-7$$

which may be rewritten as $$y = \int (5x^3+2x-7)\,dx$$

∴ $$y = \frac{5}{4}x^4+x^2-7x+C$$

This is the required general solution.

(b) To find the particular solution we must substitute $y = 2$ when $x = 1$ into the general solution

∴ $$2 = \frac{5}{4}(1)^4+(1)^2-7(1)+C$$

from which $$C = 2-\frac{5}{4}-1+7$$

$$= 6.75$$

Hence the required particular solution is $y = \dfrac{5}{4}x^4+x^2-7x+6.75.$

Exercise 16

1) Sketch the family of curves represented by the differential equation $\dfrac{dy}{dx} = x,$ and find the equation of the curve passing through the point (2, 3).

2) Sketch the family of curves represented by the differential equation $\dfrac{dy}{dx} = 3x^2$. Find also the equation of the curve which passes through the point $(5, -3)$.

3) Find the general solution of the differential equation $\dfrac{dy}{dx} = x^2 - 5x$ and the particular solution if $x = 4$ when $y = 0$.

4) If $\dfrac{dy}{dx} = 6x^3 + 5x^2 + 7$ find the particular solution if it represents the equation of the curve which passes through the point $(2, 2)$.

5) If $\dfrac{dy}{dx} = ax$, where a is a constant, find the particular solution if $x = 0$ when $y = 0$, and also $x = 2$ when $y = 4$.

EQUATIONS OF THE TYPE $\frac{dy}{dx} = ky$

We will show now that $y = Ae^{kx}$ is the general solution of the differential equation $\dfrac{dy}{dx} = ky$. In order to do this it is necessary to prove that $y = Ae^{kx}$ satisfies the equation $\dfrac{dy}{dx} = ky$.

Now
$$y = Ae^{kx} \qquad [1]$$

and differentiating with respect to x we have $\dfrac{dy}{dx} = Ake^{kx}$ $\qquad [2]$

Now for the given differential equation:

$$\text{the L.H.S.} = \frac{dy}{dx}$$

$$= Ake^{kx} \quad \text{from equation [2]}$$

$$= k(Ae^{kx})$$

$$= ky \qquad \text{from equation [1]}$$

$$= \text{R.H.S.}$$

Hence the general solution of $\dfrac{dy}{dx} = ky$ is $y = Ae^{kx}$, where A is a constant.

EXAMPLE 4

Given the differential equation $\dfrac{dy}{dx} = 6y$ find:

(a)   the general solution, and

(b)   the particular solution if $x = 0$ when $y = 3$.

(a)   Here the equation is of the form $\dfrac{dy}{dx} = ky$ where $k = 6$.
Hence the general solution is $y = Ae^{6x}$.

(b)   Substituting   the   values $x = 0$ and $y = 3$ into   the   general
solution we have $3 = Ae^{6 \times 0}$

from which $A = 3$ since $e^0 = 1$

Hence the particular solution is $y = 3e^{6x}$.

EXAMPLE 5

In a particular engineering problem the rate of change of distance s with
respect to time t is known to be proportional to the distance s. Express
this as a differential equation and find:

(a)   the   particular   solution   if $s = 2$ when $t = 1$,   and $s = 6$
when $t = 1.5$.

(b)   the value of s when $t = 3$.

In mathematical notation the rate of change of s with respect to t is $\dfrac{ds}{dt}$.

Now we are given $\dfrac{ds}{dt}$ is proportional to s,

that is $\dfrac{ds}{dt} = ks,$ where k is a constant.

The general solution of this differential equation is $s = Ae^{kt}$ where both A
and k are constants. The boundary conditions given in part (a) will enable
us to find the values of these two constants.

(a)   Now $s = 2$ when $t = 1$,   and substituting these values into the
general solution we have $2 = Ae^{k \times 1}$ [1]

Similarly, since $s = 6$ when $t = 1.5$ then $6 = Ae^{k \times 1.5}$ [2]

Dividing equation [2] by equation [1] then

$$\frac{6}{2} = \frac{e^{1.5k}}{e^k}$$

∴ $$3 = e^{(1.5-1)k}$$

or $$3 = e^{0.5k}$$

To solve this equation for k we must put it into logarithmic form,

that is $$0.5k = \log_e 3$$

∴ $$k = \frac{1.10}{0.5}$$

∴ $$k = 2.20$$

To find A we will substitute $k = 2.20$ into equation [1]

∴ $$2 = Ae^{2.20}$$

or $$A = \frac{2}{9.03}$$

∴ $$A = 0.221$$

Hence the required solution is $s = 0.221e^{2.20t}$

(b) When $t = 3$ the corresponding value of s may be found by substituting $t = 3$ into the particular solution

∴ $$s = 0.221e^{2.20 \times 3}$$

$$= 162$$

EXAMPLE 6

Fig. 8.3 shows the general arrangement of a belt drive.

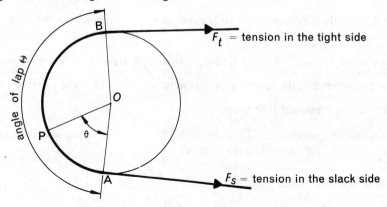

Fig. 8.3

It can be shown that the rate of change of the tension F N with respect to the angle θ radian at any point P is given by μF, where μ is the coefficient of friction between the belt and the pulley.

Find the equation connecting F_t, F_s, μ, and Θ' and hence find the value of F_t when $F_s = 50$ N, $\mu = 0.3$ and the angle of lap is $120°$.

The initial differential equation will be $\qquad \dfrac{dF}{d\theta} = \mu F$

and hence the general solution will be of the form $\qquad F = Ae^{\mu\theta}$

Now, when $\theta = \Theta$, $F = F_s$ $\therefore$ $\qquad\qquad F_s = Ae^{\mu \times 0}$

and since $\quad e^0 = 1 \qquad\qquad$ then $\qquad\qquad A = F_s$

and hence the solution becomes $\qquad\qquad F = F_s e^{\mu\theta}$

Also, when $\theta = \Theta'$, $F = F_t$ $\therefore$ $\qquad\qquad F_t = F_s e^{\mu\Theta'}$

Hence the required equation is $\quad F_t = F_s e^{\mu\Theta'}$.

As in many mathematical equations the angle must be expressed in radians.

Since $\qquad\qquad\qquad 360° = 2\pi$ radians

then $\qquad\qquad\qquad 120° = \dfrac{2}{3}\pi = 2.09$ radians.

Hence substituting the given values into the solution we have

$$F_t = 50e^{0.3 \times 2.09}$$

$$= 93.6 \text{ N}$$

EXAMPLE 7

Newton's law of cooling states that the rate at which the temperature T of a body falls is proportional to the difference in temperature between the body and its surroundings.

If t is the time and the surroundings are at 0 °C then the differential equation representing the above information is $\dfrac{dT}{dt} = -kT$, the negative sign indicating a temperature drop.

If a body's temperature falls from 90 °C to 70 °C in 50 seconds find how long it will take to cool another 20 °C.

Since the differential equation given is $\qquad \dfrac{dT}{dt} = -kT$

then the general solution is of the form $\qquad T = Ae^{-kt}$

If we assume that cooling starts at 90 °C then $T = 90$ when $t = 0$ s and substituting these values into the general equation we have:

$$90 = Ae^{-k \times 0}$$

and since $e^0 = 1$ then $A = 90$

Also, when $T = 70$ °C then $t = 50$ s and substituting these values into $T = 90e^{-kt}$ we get $70 = 90e^{-k \times 50}$

$\therefore$ $e^{50k} = \dfrac{90}{70}$

and rearranging into logarithmic form $50k = \log_e\left(\dfrac{90}{70}\right)$

$\therefore$ $k = 0.00503$

hence the particular solution is $T = 90e^{-0.00503t}$

We shall now find the time to cool another 20 °C, that is to 50 °C. If we substitute $T = 50$ °C into $T = 90e^{-0.00503t}$ the value of t will be the time to cool from 90 °C, that is when the time commences, to 50 °C.

$\therefore$ substituting $50 = 90e^{-0.00503t}$

$$50 = \dfrac{90}{e^{0.00503t}}$$

$$e^{0.00503t} = \dfrac{90}{50}$$

and rearranging in logarithmic form:

$$0.00503t = \log_e\left(\dfrac{90}{50}\right)$$

$\therefore$ $t = 117$ s

Hence the time to cool from 70 °C to 50 °C $= 117 - 50$

$$= 67 \text{ seconds}$$

Exercise 17

1) Find the general solution of the differential equation $\dfrac{dy}{dx} = 3y$ and

the particular solution if $x = 0$ when $y = 2$. Find also the value of y when $x = 2$.

2)　Find the particular solution of the differential equation $\dfrac{ds}{dt} = -ks$ if $s = 7$ when $t = 1$, and also $s = 4$ when $t = 2$. Find also the value of t when $s = 2$.

3)　The rate at which the current I A dies in an electrical circuit which contains a resistance $R\,\Omega$ and an inductance L H is given by the differential equation $\dfrac{dI}{dt} = -\dfrac{R}{L}I$ where t s is time. If $R = 2\,\Omega$ and $L = 0.06$ H, and $I = 10$ A when $t = 0$ s, find the solution of the differential equation. Find also the current after 0.02 s.

4)　A radioactive material decays at a rate which is proportional to the amount of radioactivity remaining. If the amount of radioactivity remaining is denoted by N and the time is t form a differential equation which represents this statement. If the half-life (that is the time taken for half of the radioactivity to decay) is 10 years, how long will it take for $\frac{3}{4}$ of the radioactivity to decay?

5)　An electrical circuit contains a resistance R ohm and a capacitor C farad which initially holds a charge Q coulomb. The rate of discharge of the capacitor is given by the equation $\dfrac{dQ}{dt} + \dfrac{Q}{RC} = 0$, where t seconds is the time. If $R = 80\,000$ ohm and $C = 0.3 \times 10^{-6}$ farad, and also $Q = 0.0015$ coulomb initially, find the equation connecting Q and t. Find Q when $t = 0.02$ seconds and also t when $Q = 0.001$ coulomb.

SUMMARY

The differential equation $\qquad \dfrac{dy}{dx} = k.y$

has a general solution of the form $\quad y = A.e^{kx}$

THREE DIMENSIONAL TRIANGULATION PROBLEMS

9.

On reaching the end of this chapter you should be able to :-

1. Solve plane triangles.
2. Define the angle between a line and a plane.
3. Define the angle between two intersecting planes.
4. Identify relevant planes in a given three dimensional problem.

5. Solve a three dimensional triangulation problem capable of being specified within a rectangular prism.

THE SOLUTION OF PLANE TRIANGLES

We have met previously the *sine* and *cosine* rules for finding the sides and angles of non-right angled triangles, and also three formulae for calculating the areas of triangles.

Fig. 9.1 shows a triangle labelled conventionally, and for convenience the formula are listed below:

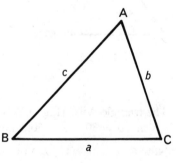

The *sine* rule: $\dfrac{a}{\sin A} = \dfrac{b}{\sin B} = \dfrac{c}{\sin C}$

This is used when given: one side and any two angles,

or: two sides and an angle opposite one of the sides.

Fig. 9.1

The *cosine* rule: $a^2 = b^2 + c^2 - 2bc.\cos A$

or: $b^2 = a^2 + c^2 - 2ac.\cos B$

or: $c^2 = a^2 + b^2 - 2ab.\cos C$

This is used when given: two sides and the angle between them

or: the three sides.

119

The *area* of a triangle may be found using:

either   Area $= \frac{1}{2} \times$ base $\times$ altitude

or   Area $= \frac{1}{2}ab.\sin C = \frac{1}{2}bc.\sin A = \frac{1}{2}ac.\sin B$

or   Area $= \sqrt{s(s-a)(s-b)(s-c)}$ where $s = \dfrac{a+b+c}{2}$

The diameter D of the circumscribing circle of a triangle is given by

$$D = \frac{a}{\sin A} = \frac{b}{\sin B} = \frac{c}{\sin C}$$

This topic was covered extensively in Level II *Mechanical Engineering Mathematics*, and the worked example which follows should be sufficient revision.

EXAMPLE 1

Find the resultant of the two forces shown in Fig. 9.2 and the angle it makes with the 50 N force.

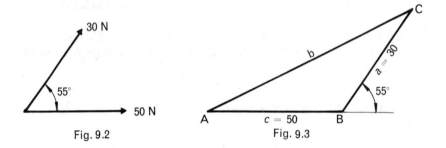

Fig. 9.2   Fig. 9.3

The triangle ABC (Fig. 9.3) is the vector diagram for the given system in which $a = 30,$ $c = 50,$ and the length b gives the resultant.

Now $\angle ABC = 180° - 55° = 125°$

To find b we will use the cosine rule which gives

$$b^2 = a^2 + c^2 - 2ac.\cos B$$

$\therefore$ $b^2 = 30^2 + 50^2 - 2 \times 30 \times 50 \times \cos 125°$

$$= 900 + 2500 - 3000(-0.5736)$$

$$= 5121$$

$\therefore$ $b = \sqrt{5121} = 71.56$

To find angle A we will use the sine rule which gives:

$$\frac{a}{\sin A} = \frac{b}{\sin B}$$

from which

$$\sin A = \frac{a(\sin B)}{b}$$

$$= \frac{30(\sin 125°)}{71.56}$$

$$= \frac{30 \times 0.8192}{71.56}$$

$$= 0.3434$$

∴ $$A = 20° 5'$$

The resultant is 71.56 N and the angle it makes with the 50 N force is 20°5′.

Exercise 18

1) The line diagram of a jib crane is shown in Fig. 9.4. Calculate the length of the jib BC and the angle between the jib BC and the tie rod AC.

2) The schematic layout of a petrol engine is shown in Fig. 9.5. The crank OC is 80 mm long and the connecting rod CP is 235 mm long. For the position shown find the distance OP and the angle between the crank and the connecting rod.

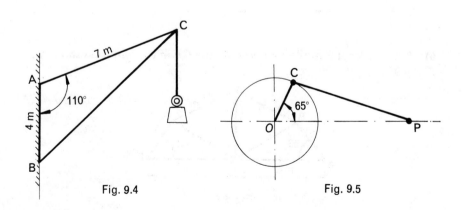

Fig. 9.4   Fig. 9.5

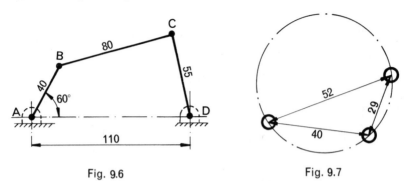

Fig. 9.6   Fig. 9.7

3) A four-bar mechanism ABCD is shown in Fig. 9.6. Links AB, BC and CD are pin jointed at their ends. (The name four-bar is given because the length AD is considered to be a link.) Find the length AC and the angle ADC.

4) Fig. 9.7 shows the dimensions between three holes which are to be drilled in a plate of steel. Find the radius of the circle on which the holes lie.

5) P, Q and R are sliders constrained to move along centre-lines OA and OB as shown in Fig. 9.8. For the given position find the distance OR.

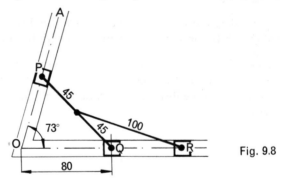

Fig. 9.8

6) Find the lengths of the members BF and CF in the roof truss shown in Fig. 9.9.

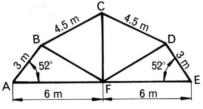

Fig. 9.9

7) Find the area of triangle ABC in Fig. 9.4.

8) Find the area of triangle OCP in Fig. 9.5.

9) Find the area of quadrilateral ABCD in Fig. 9.6.

10) Find the area of the triangle in Fig. 9.7.

THREE DIMENSIONAL TRIANGULATION PROBLEMS

Since most components, assemblies, structures, etc. in engineering are three dimensional it is inevitable that we will have to solve problems of this type. These problems may involve finding angles between lines, true lengths of lines, and even the angle between two planes.

This topic is also useful because it helps us visualise problems in three dimensions, which is an asset for all good engineers.

Most three dimensional triangulation problems may be solved using only basic trigonometrical ratios and the theorem of Pythagoras for suitably selected right angled triangles.

No specific method can be laid down for these problems but we feel that, after you have worked through the examples in the text, you will be able to approach with confidence the exercises at the end of the chapter.

EXAMPLE 2

Find angle BAC in the triangular prism shown in Fig. 9.10.

On inspection it may be seen that several of the faces of the prism are right angled triangles. Some of these triangles are used in the following solution.

Using the right angled triangle OAB

we have, $\sin 47° = \dfrac{AB}{OB}$

∴ $AB = OB \sin 47°$

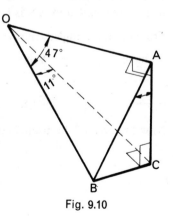

Fig. 9.10

Using the right angled triangle OCB

we have, $\sin 11° = \dfrac{BC}{OB}$

$\therefore$ $BC = OB \sin 11°$

Using the right angled triangle ABC

we have, $\sin B\hat{A}C = \dfrac{BC}{AB} = \dfrac{OB \sin 11°}{OB \sin 47°} = \dfrac{\sin 11°}{\sin 47°}$

hence $B\hat{A}C = 15° \, 8'$

EXAMPLE 3

Fig. 9.11 shows a component which has a rectangular base. The top edge is parallel to the base and one of the triangular end faces is perpendicular to the base. Find the area of the sloping triangular end face.

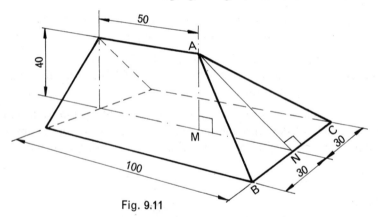

Fig. 9.11

Two construction lines have been drawn. AM is perpendicular to the rectangular base and AN is perpendicular to the edge BC.

From the right angled triangle AMN and using the theorem of Pythagoras

we have, $AN^2 = AM^2 + MN^2$

$= 40^2 + 50^2 = 4100$

$\therefore$ $AN = 64.03 \text{ mm}$

Now the area of the triangular sloping face ABC:

$= \tfrac{1}{2} \times BC \times AN$

$= \tfrac{1}{2} \times 60 \times 64.03$

$= 1921 \text{ mm}^2$

THE ANGLE BETWEEN A LINE AND A PLANE

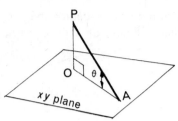

In Fig. 9.12 the line AP intersects the *xy* plane at A. To find the angle between AP and the plane, draw PM perpendicular to the plane and then join AM.

The angle between the line and the plane is ∠PAM.

Fig. 9.12

EXAMPLE 4

Find the angle θ in the rectangular block shown in Fig. 9.13.

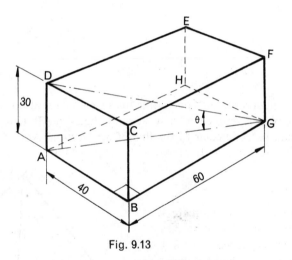

Fig. 9.13

From the right angled triangle ABG and using the theorem of Pythagoras we have,

$$AG^2 = AB^2 + BG^2$$

$$= 40^2 + 60^2 = 5200$$

∴ $$AG = 72.11 \text{ mm}$$

In right angled triangle AGD

we have, $$\tan \theta = \frac{AD}{AG} = \frac{30}{72.11} = 0.4160$$

∴ $$\theta = 22°35'$$

THE ANGLE BETWEEN TWO PLANES

The problem of finding the angle between two planes resolves itself into that of finding an angle between two lines, as shown in Fig. 9.14.

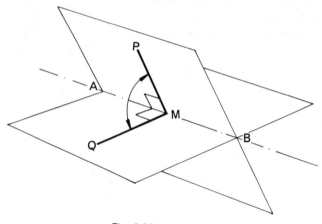

Fig. 9.14

Each of the planes contains one of the lines which meet at the intersection, AB, of the planes. Also each line is at right angles to AB, that is both PM and QM are perpendicular to the line of intersection, AB, of the planes. The true angle between the planes is the angle PMQ.

EXAMPLE 5

A right pyramid has a square base 60 mm square, and a height of 50 mm. Find (a) the angle between the base and a sloping face, and (b) the angle between two adjacent sloping faces.

The pyramid is shown suitably labelled in Fig. 9.15.

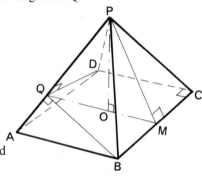

Fig. 9.15

(a) BC is the line of intersection of a sloping side and the base. PO is the vertical centre-line. If OM is drawn perpendicular to BC, then PM is also perpendicular to BC. Hence the angle PMO is the required angle between the base and a sloping side.

Now using the right angled triangle POM

we have, $\tan P\hat{M}O = \dfrac{PO}{OM} = \dfrac{50}{30} = 1.667$

$\therefore$ $P\hat{M}O = 59° \; 2'$

(b) AP is the line of intersection of two adjacent sloping faces. BQ is drawn perpendicular to AP. Then DQ is also perpendicular to AP. Hence angle BQD is the required angle between two adjacent sloping faces.

Now using the right angled triangle POM and the theorem of Pythagoras we have,

$$PM^2 = PO^2 + OM^2 = 50^2 + 30^2 = 3400$$

$\therefore$ $PM = 58.31 \text{ mm}$

In right angled triangle PMB

we have, $\tan P\hat{B}M = \dfrac{PM}{BM} = \dfrac{58.31}{30} = 1.944$

$\therefore$ $P\hat{B}M = 62° \; 47'$

From the symmetry of the figure $Q\hat{A}B = P\hat{B}M = 62° \; 47'$

In right angled triangle AQB

we have, $QB = AB \sin Q\hat{A}B$

$= 60 \sin 62° \; 47'$

$= 53.36 \text{ mm}$

From the right angled triangle BCD and using the theorem of Pythagoras we have,

$$BD^2 = DC^2 + BC^2 = 60^2 + 60^2 = 7200$$

$\therefore$ $BD = 84.85 \text{ mm}$

Hence $BO = \frac{1}{2} BD = \frac{1}{2}(84.85) = 42.43 \text{ mm}$

From the right angled triangle BOQ

we have, $\sin B\hat{Q}O = \dfrac{BO}{QB} = \dfrac{42.43}{53.36} = 0.7952$

$\therefore$ $B\hat{Q}O = 52° \; 40'$

But $B\hat{Q}D = 2 \,(BQO) = 2 \,(52° \; 40')$

$\therefore$ $B\hat{Q}D = 105°20'$

EXAMPLE 6

A triangular prism OABC is cut off from a rectangular block by an oblique plane OAC as shown in Fig. 9.16.

Find (a) the angle between the oblique plane OAC and the top surface of the block, and (b) the angle DOC where AD is perpendicular to the top face and CD is perpendicular to the front face of the block.

(a)   Draw BM perpendicular to AO and then join CM which will also be perpendicular to AO. Since both these lines are perpendicular to the line of intersection AO of the plane OAC and the top face of the block, then angle CMB is the required angle.

Again we shall use several right angles triangles to find this angle.

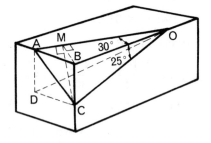

Fig. 9.16

Using the right angled triangle BMO

we have, $\sin 30° = \dfrac{BM}{OB}$

$\therefore$ $BM = OB \sin 30°$

Using the right angled triangle OBC

we have, $\tan 25° = \dfrac{BC}{OB}$

$\therefore$ $BC = OB \tan 25°$

Using the right angled triangle CMB

we have, $\tan C\hat{M}B = \dfrac{BC}{BM} = \dfrac{OB \tan 25°}{OB \sin 30°} = \dfrac{\tan 25°}{\sin 30°}$

$\therefore$ $C\hat{M}B = 43°$

(b) Using the right angled triangle ABO

we have, $\tan 30° = \dfrac{AB}{OB}$

$\therefore$ $AB = OB \tan 30°$

but ABCD is a rectangle and hence $DC = AB$

$\therefore$ $DC = AB = OB \tan 30°$

Using the right angled triangle OBC

we have, $\cos 25° = \dfrac{OB}{OC}$

$\therefore$ $OC = OB \sec 25°$

Using the right angled triangle OCD

we have, $\tan C\hat{O}D = \dfrac{DC}{CO} = \dfrac{OB \tan 30°}{OB \sec 25°} = (\tan 30°)(\cos 25°)$

$\therefore$ $C\hat{O}D = 27° \, 37'$

Exercise 19

1) Fig. 9.17 shows a small component which has a horizontal rectangular base ABCD in which AD is 70 mm and CD is 40 mm. The end face of the component AXB is vertical and it is an equilateral triangle. The top edge XY is horizontal and it is 45 mm long. Find:

(a) The altitude of $\triangle$YDC (YF in the diagram).

(b) The area of the sloping face YDC.

(c) The angle which the sloping face YDC makes with the base.

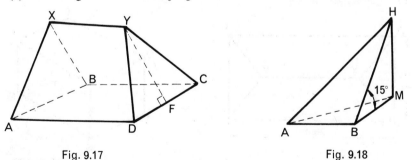

Fig. 9.17   Fig. 9.18

2) Fig. 9.18 shows a triangular prism in which HM is vertical and face ABM is horizontal. The edge HB slopes at 15° to the horizontal. If AB is 55 mm, $\angle HBA = 90°$ and $\angle HAB = 55°$,   find:

(a)   The length HB.

(b)   The vertical height HM.

(c)   The angle HAM which AH makes with the horizontal.

3)   The base of the wedge shown in Fig. 9.19 is a rectangle 80 mm long and 60 mm wide. The vertical faces ABC and PQR are equilateral triangles. Calculate:

(a)   The angle between the diagonals PB and PC.

(b)   The angle between the diagonal PC and the base.

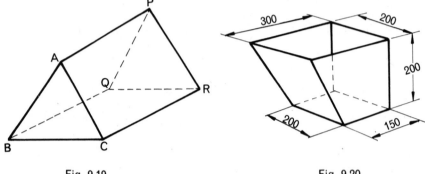

Fig. 9.19   Fig. 9.20

4)   Fig. 9.20 shows an elevator hopper with a square back. Find the area of metal required to make it.

5)   Fig. 9.21 shows, in plan view, the roof of a rectangular building, the sloping faces of which are inclined at 34° to the horizontal. Calculate:

(a)   The height of the ridge XY above the horizontal plane ABCD.

(b)   The angle which the edge AX makes with the horizontal.

(c)   The length of AX.

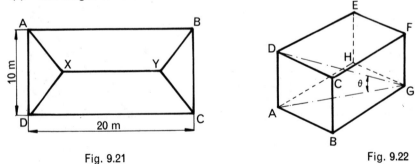

Fig. 9.21   Fig. 9.22

6)   Fig. 9.22 shows a rectangular block 70 mm long, 50 mm wide and 40 mm high. Find the lengths of AG and GD and the angle θ.

7)   The component shown in Fig. 9.23 has a square base and each of the sloping faces makes an angle of 30° with the vertical. Calculate:

(a)   The angle between the edge AB and the base of the block.

(b)   The true angle between two adjacent sloping faces.

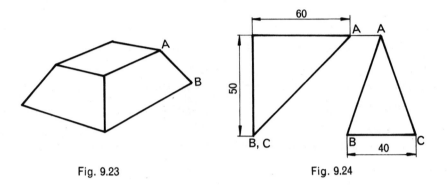

Fig. 9.23   Fig. 9.24

8)   Two orthographic third angle views of a symmetrical triangular prism are shown in Fig. 9.24. Calculate the true angle BAC.

10. TRIGONOMETRICAL GRAPHS

On reaching the end of this chapter you should be able to:-

1. Define angular velocity ω radians per second.

2. Define periodic time as $\dfrac{2\pi}{\omega}$

3. Define frequency.

4. Plot, on the same graph, the curves of sin ωt

and cos ωt for ωt, ω = ½, 1, 2, and 3, and values of t in seconds.

5. Plot the curves of sin² ωt and cos² ωt for more than one cycle.

6. Plot R sin(ωt±α) for given values of R, ω, and α.

7. Interpret α as a phase lead or phase lag.

AMPLITUDE OR PEAK VALUE

The graphs of sin θ and cos θ each have a maximum value of $+1$ and a minimum value of -1.

Similarly the graphs of $R.\sin\theta$ and $R.\cos\theta$ each have a maximum value of $+R$ and a minimum value of $-R$. These graphs are shown in Fig. 10.1.

The value of R is known as the amplitude or peak value.

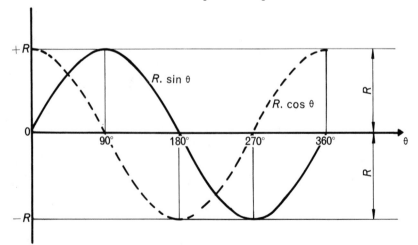

Fig. 10.1

EXAMPLE 1

Draw the curve of $E = 100 \sin \theta$.

Using the table of sines we draw up the table shown below.

$\theta°$	0	30	60	90	120	150	180
$\sin \theta$	0	0.500 0	0.866 0	1.000	0.866 0	0.500 0	0
$100 \sin \theta$	0	50	86.6	100	86.6	50	0

$\theta°$	210	240	270	300	330	360
$\sin \theta$	−0.500 0	−0.866 0	−1.000	−0.866 0	−0.500 0	0
$100 \sin \theta$	−50	−86.6	−100	−86.6	−50	0

The curve is shown in Fig. 10.2.

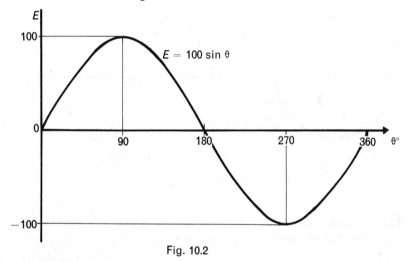

Fig. 10.2

RELATION BETWEEN ANGULAR AND TIME SCALES

In Fig. 10.3 OP represents a radius, of length R, which rotates at a uniform angular velocity ω radians per second about O, the direction of rotation being anticlockwise.

Now   angular velocity $= \dfrac{\text{angle turned through}}{\text{time taken}}$

∴   angle turned through $=$ (angular velocity) $\times$ (time taken)

and hence after a time t seconds

          angle turned through $= \omega t$ radians

Also from the right angled triangle OPM:

$$\frac{PM}{OP} = \sin P\hat{O}M$$

$\therefore$ $PM = O$P.$\sin P\hat{O}M$

or $PM = R.\sin \omega t$

If a graph is drawn, as in Fig. 10.3, showing how PM varies with the angle ωt the sine wave representing $R.\sin \omega t$ is obtained. It can be seen that the peak value of this sine wave is R (i.e. the magnitude of the rotating radius).

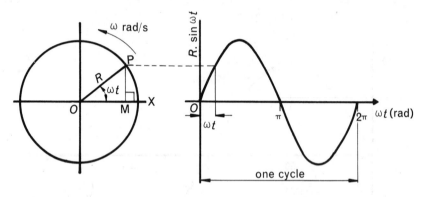

Fig. 10.3

The horizontal scale shows the angle turned through, ωt, and the wave-form is said to be plotted on an **angular** or ωt **base**.

CYCLE

A cycle is the portion of the waveform which shows its complete shape without any repetition. It may be seen from Fig. 10.3 that one cycle is completed whilst the radius OP turns through $360°$ or 2π radians.

PERIOD

This is the time taken for the waveform to complete one cycle.

It will also be the time taken for OP to complete one revolution or 2π radians.

Now we know that   time taken $= \dfrac{\text{angle turned through}}{\text{angular velocity}}$

hence

$$\boxed{\text{the period} = \dfrac{2\pi}{\omega} \text{ seconds}}$$

FREQUENCY

The number of cycles per second is called the frequency. The unit of frequency representing one cycle per second is the hertz (Hz).

Now if   1 cycle is completed in $\dfrac{2\pi}{\omega}$ seconds (a period)

then $1 \div \dfrac{2\pi}{\omega}$ cycles are completed in 1 second

and therefore $\dfrac{\omega}{2\pi}$ cycles are completed in 1 second

Hence

$$\boxed{\text{frequency} = \dfrac{\omega}{2\pi} \text{ Hz}}$$

Also since $\text{period} = \dfrac{2\pi}{\omega} \text{ s}$

then

$$\boxed{\text{frequency} = \dfrac{1}{\text{period}}}$$

TIME BASE

We have seen how a graph may be plotted on an 'angular' or 'ωt' base as in Fig. 10.3. Alternatively the units on the horizontal axis may be those of time (usually seconds), and this is called a 'time' base.

RADIANS AND DEGREES

We know that one full revolution is equivalent to 360° or 2π radians.

Hence $1 \text{ radian} = \left(\dfrac{360}{2\pi}\right)^{\circ} = \left(\dfrac{180}{\pi}\right)^{\circ}$

If tables are used when finding trigonometrical ratios, such as sines and cosines, of angles it may be necessary to convert an angle in radians to an angle in degrees.

For example $\quad \sin 0.5 = \sin\left(0.5 \times \frac{180}{\pi}\right)^{\circ} = \sin 28.65^{\circ} = 0.4795$

It should be noted that if the units of an angle are omitted it is assumed that it is given in radians (as in the above example).

If a scientific calculator is available it is often possible to set the machine to accept radians by setting a special key. There is then no necessity to convert from radians to degrees.

GRAPHS OF sin *t*, sin 2*t*, sin 3*t*, AND sin ½*t*

Consider sin *t*.   Since the period of $\sin \omega t$ is $\dfrac{2\pi}{\omega}$ seconds

then   the period of sin *t* is $\dfrac{2\pi}{1} = 6.28$ seconds.

In order to plot one complete cycle of the waveform it is necessary to take values of *t* from 0 to 6.28 seconds. The reader may find it useful to draw up a suitable table of values and plot the curve. The curve is shown plotted on a time base in Fig. 10.4.

Similarly, the waveform **sin 2*t*** has a period of $\dfrac{2\pi}{2} = 3.14$ seconds

and   the waveform **sin 3*t*** has a period of $\dfrac{2\pi}{3} = 2.09$ seconds

and   the waveform **sin ½*t*** has a period of $\dfrac{2\pi}{\frac{1}{2}} = 12.56$ seconds

All these curves are shown plotted in Fig. 10.4. This enables a visual comparison to be made and it may be seen, for example, that the curve of sin 3*t* has a frequency of three times that of sin *t* (since three cycles of sin 3*t* are completed during one cycle of sin *t*).

GRAPHS OF R.cos ω*t*

The waveforms represented by *R*.cos ω*t* are similar to sine waveforms, *R* being the peak value and $\dfrac{2\pi}{\omega}$ the period. The reader is left to plot these as instructed in Exercise 21 which follows this text.

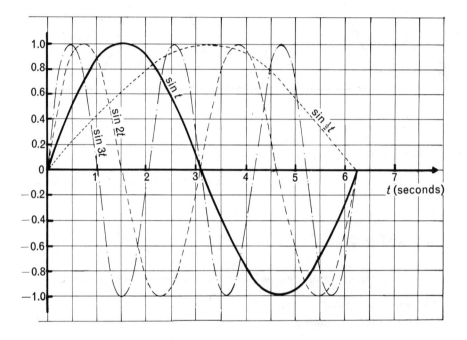

Fig. 10.4

GRAPHS OF sin²*t* AND cos²*t*

It is sometimes necessary in engineering applications, such as when finding the root mean square value of alternating currents and voltages, to be familiar with the curves sin²*t* and cos²*t*.

As previously the period of sin *t* is $\dfrac{2\pi}{1} = 6.28$ seconds, and so we will plot the graph of sin²*t* on a time base using values of *t* from 0 to 6.28 seconds. Again it may be useful for the reader to construct a suitable table of values and the resulting graph is shown in Fig. 10.5.

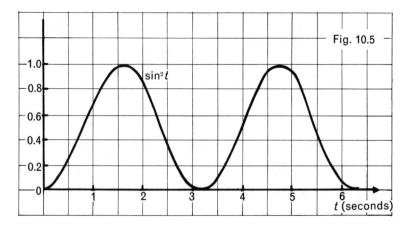

Fig. 10.5

It may be seen that the cycle of the $\sin^2 t$ curve is one half that of $\sin t$, and also the curve is wholly positive.

The reader is left to plot the $\cos^2 t$ curve which is Question 5 of Exercise 2.

Exercise 20

1) Draw the curve of $50.\cos \theta$ over one cycle on an angle base.

2) Draw the curve of $25.\sin 2\theta$ over two cycles on the same axes that were used for Question 1.

3) Find the period of the waveform $\cos t$ and then draw the curve over one cycle on a time base.

4) Draw the waveform represented by $\cos 2t$, $\cos 3t$ and $\cos \frac{1}{2}t$ on the same axes as used for Question 3 and over the period of $\cos t$.

5) Draw the curve of $\cos^2 t$ on a time base over the period of $\cos t$.

EXAMPLE 2

Draw the graph of $5.\sin (2t-1.2)$ over one cycle on a time base.

Now the time for one cycle, that is the period, for $\sin \omega t$ is $\dfrac{2\pi}{\omega}$

∴   the time for one cycle, that is the period,

for $5.\sin (2t-1.2)$ is $\dfrac{2\pi}{2} = 3.14$ seconds.

It should be noted that the period of sin $(2t-1.2)$ is the same as the period of sin $2t$. As you will see later the figure '-1.2' does not affect the period or frequency of the curve.

Hence the table of values will be from $t = 0$ to $t = 3.14$ seconds and we have chosen 0.2 second intervals.

t	0	0.2	0.4	0.6	0.8	1.0
$2t-1.2$	-1.2	-0.8	-0.4	0	0.4	0.8
sin $(2t-1.2)$	-0.932	-0.717	-0.389	0	0.389	0.717
5.sin $(2t-1.2)$	-4.66	-3.59	-1.95	0	1.95	3.59

t	1.2	1.4	1.6	1.8	2.0
$2t-1.2$	1.2	1.6	2.0	2.4	2.8
sin $(2t-1.2)$	0.932	0.999	0.909	0.675	0.335
5.sin $(2t-1.2)$	4.66	4.99	4.55	3.38	1.67

t	2.2	2.4	2.6	2.8	3.0
$2t-1.2$	3.2	3.6	4.0	4.4	4.8
sin $(2t-1.2)$	-0.058	-0.443	-0.757	-0.952	-0.996
5.sin $(2t-1.2)$	-0.292	-2.21	-3.78	-4.76	-4.98

In addition to the above values it is worth remembering that the curve cuts the axis when angle $(2t-1.2) = 0$ and when $(2t-1.2) = 2\pi$

i.e. when, $t = 0.6$ and when $t = 3.74$

It also cuts the axis half way between these two values

i.e. when, $t = \dfrac{0.6+3.74}{2} = 2.17$

The maximum peak value $+5$ will occur when the value of t is half way between $t = 0.6$ and $t = 2.17$

i.e. when, $t = \dfrac{0.6+2.17}{2} = 1.38$

The minimum peak value -5 will occur when the value of t is half way between $t = 2.17$ and $t = 3.74$

i.e. when, $t = \dfrac{2.17+3.74}{2} = 2.95$

The curve is shown plotted in Fig. 10.6.

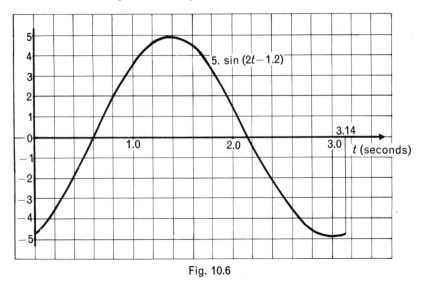

Fig. 10.6

Exercise 21

1) Draw the graph of $\sin (t-0.4)$ over one cycle on a time base.

2) Draw the graph of $\sin (t+0.7)$ over one cycle on a time base.

3) Find the period of the waveform $3.\sin (4t-2)$ and draw the graph over one cycle on a time base.

4) Plot the curve $4.\sin (3t+1.8)$ over one cycle, and find the values of the peak value and the period. Use a time base.

5) Find the equation of the waveform which has a frequency which is twice that of the curve $\sin t$ and has an amplitude of 3. Draw these two curves on the same axes on a time base over the period of $\sin t$.

PHASE ANGLE

The principal use of sine and cosine waveforms occurs in electrical engineering where they represent alternating currents and voltages. In a diagram such as shown in Fig. 10.7 the rotating radii OP and OQ are called phasors.

Fig. 10.7 shows two phasors OP and OQ, separated by an angle α, rotating at the same angular speed in an anti-clockwise direction. The sine waves produced by OP and OQ are identical curves but they are displaced from each other. The amount of displacement is known as the phase difference and, measured along the horizontal axis, is α. The angle α is called the **phase angle**.

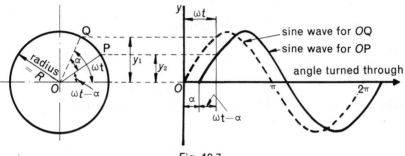

Fig. 10.7

In Fig. 10.7 the phasor OP is said to *lag* behind phasor OQ by the angle α. If the radius of the phasor circle is R then $OP = OQ = R$ and hence:

for the phasor OQ, $y_2 = R.\sin \omega t$

and for the phasor OP, $y_1 = R.\sin (\omega t - \alpha)$

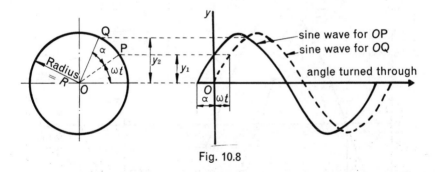

Fig. 10.8

Similarly in Fig. 10.8 the phasor OP leads the phasor OQ by the phase angle, α.

Hence for the phasor OQ, $y_1 = R.\sin \omega t$

and for the phasor OP, $y_2 = R.\sin (\omega t + \alpha)$

In practice it is usual to draw waveform on an 'angular' or 'ωt' bases, when considering phase angles as in the following example.

EXAMPLE 3

Plot the waveform of $\sin \omega t$ and $\sin\left(\omega t - \dfrac{\pi}{3}\right)$ on an angular base and identify the phase angle.

The cycle of a sine wave is 360° or 2π radians.

Hence $\sin \omega t$ will be plotted between when $\omega t = 0$ and when $\omega t = 2\pi$ radians.

Also $\sin\left(\omega t - \dfrac{\pi}{3}\right)$ will be plotted between values given by:

$$\omega t - \frac{\pi}{3} = 0 \quad \text{and when} \quad \omega t - \frac{\pi}{3} = 2\pi$$

i.e. $$\omega t = \frac{\pi}{3} \text{ radians and} \quad \omega t = 2\pi + \frac{\pi}{3} = \frac{7\pi}{3} \text{ radians}$$

The table of values have been omitted, but the reader may find it useful to check these values, and hence the graphs which are shown plotted in Fig. 10.9.

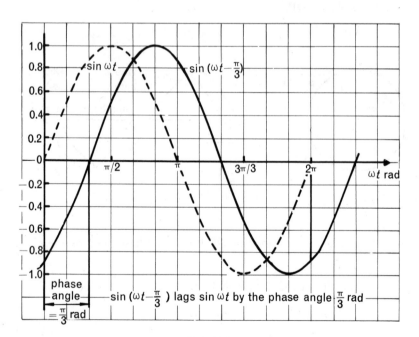

Fig. 10.9

Exercise 22

1) Plot the graphs of $\sin\theta$ and $\sin(\theta+0.9)$ on the same axes on an angular base using units in radians. Indicate the phase angle between the waveforms and explain whether it is an angle of lead or lag.

2) Plot the graph of $\sin\left(\omega t+\dfrac{\pi}{6}\right)$ on an angular base over one cycle.

3) Plot the graphs of $\sin\left(\omega t+\dfrac{\pi}{3}\right)$ and $\sin\left(\omega t-\dfrac{\pi}{4}\right)$ on the same axes on an angular base showing a cycle of each waveform. Identify the phase angle between the curves.

4) Write down the equation of the waveform which:

(a) leads $\sin\omega t$ by $\dfrac{\pi}{2}$ radians.

(b) lags $\sin\omega t$ by π radians.

(c) leads $\sin\left(\omega t-\dfrac{\pi}{3}\right)$ by $\dfrac{\pi}{3}$ radians.

(d) lags $\sin\left(\omega t+\dfrac{\pi}{6}\right)$ by $\dfrac{\pi}{3}$ radians.

SUMMARY

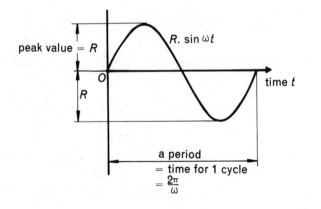

peak value $= R$

$R.\sin\omega t$

time t

a period
= time for 1 cycle
$= \dfrac{2\pi}{\omega}$

$$\text{Frequency} = \frac{1}{\text{period}} = \frac{\omega}{2\pi}$$

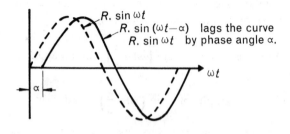

R. sin ωt
R. sin $(\omega t - \alpha)$ lags the curve
R. sin ωt by phase angle α.

ωt

α

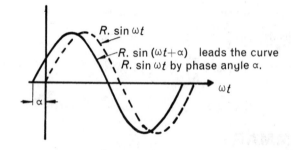

R. sin ωt
R. sin $(\omega t + \alpha)$ leads the curve
R. sin ωt by phase angle α.

ωt

α

COMPOUND ANGLE FORMULAE

On reaching the end of this chapter you should be able to :-

1. Use the formulae
   $\sin(A\pm B) = \sin A \cos B \pm \cos A \sin B$ and
   $\cos(A\pm B) = \cos A \cos B \pm \sin A \sin B.$

2. Express $R\sin(\omega t \pm \alpha)$ in the form
   $a\sin\omega t \pm b\cos\omega t$ using 1 and vice versa.
3. Deduce the relationship between a, b, R, and α.

It can be shown that:

$$\sin(A+B) = \sin A.\cos B + \cos A.\sin B$$

$$\sin(A-B) = \sin A.\cos B - \cos A.\sin B$$

$$\cos(A+B) = \cos A.\cos B - \sin A.\sin B$$

$$\cos(A-B) = \cos A.\cos B + \sin A.\sin B$$

On inspecting these formulae you may feel that an error has been made in the 'signs' on the right hand sides of the equations for $\cos(A\pm B)$. They are correct, however, and a special note should be made of them.

Since the above formulae involve the two angles A and B they are called *compound angle* formulae.

The following examples show some uses of the above formulae.

EXAMPLE 1

Simplify   (a) $\sin(\theta+90°)$ (b) $\cos(\theta-270°)$.

(a)   Using $\sin(A+B) = \sin A.\cos B + \cos A.\sin B$

and substituting θ for A and $90°$ for B

we have $\sin(\theta+90°) = \sin\theta.\cos 90° + \cos\theta.\sin 90°$

$$= (\sin\theta)0 + (\cos\theta)1$$

$$= \cos\theta$$

(b)   Using $\cos (A-B) = \cos A.\cos B + \sin A.\sin B$

and substituting θ for A and 270° for B

we have $\cos (\theta - 270°) = \cos \theta.\cos 270° + \sin \theta.\sin 270°$

$$= (\cos \theta)0 + (\sin \theta)(-1)$$

$$= -\sin \theta$$

EXAMPLE 2

By using the formula for $\cos (A+B)$ find a formula for $\cos 2\theta$.

We have $\cos (A+B) = \cos A.\cos B - \sin A.\sin B$

and substituting θ for both A and B

then $\cos (\theta + \theta) = \cos \theta.\cos \theta - \sin \theta.\sin \theta$

$\therefore$ $\cos 2\theta = \cos^2\theta - \sin^2\theta$

EXAMPLE 3

By using the sines and cosines of 30° and 45° find sin 75° and cos 15°.

To find the trigonometrical ratios of 30° and 45° we use suitable right angled triangles shown in Figs. 11.1 and 11.2.

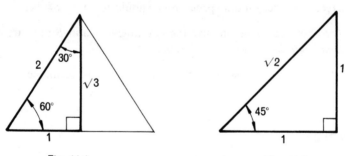

Fig. 11.1   Fig. 11.2

From Fig. 11.1 we have $\sin 30° = \dfrac{1}{2}$ and $\cos 30° = \dfrac{\sqrt{3}}{2}$

and from Fig. 11.2 we have $\sin 45° = \dfrac{1}{\sqrt{2}}$ and $\cos 45° = \dfrac{1}{\sqrt{2}}$

Now $\sin (A+B) = \sin A.\cos B + \cos A.\sin B$

and if we substitute 45° for A and 30° for B

then $\qquad$ $\sin (45°+30°) = \sin 45°.\cos 30°+\cos 45°.\sin 30°$

$\therefore$ $\qquad\qquad$ $\sin 75° = \dfrac{1}{\sqrt{2}}.\dfrac{\sqrt{3}}{2}+\dfrac{1}{\sqrt{2}}.\dfrac{1}{2}$

$\qquad\qquad\qquad\qquad = \dfrac{\sqrt{3}+1}{2\sqrt{2}}$

$\qquad\qquad\qquad\qquad = 0.966$

Now $\qquad\qquad$ $\cos (A-B) = \cos A.\cos B+\sin A.\sin B$

and if we substitute $45°$ for A and $30°$ for B

then $\qquad$ $\cos (45°-30°) = \cos 45°.\cos 30°+\sin 45°.\sin 30°$

$\therefore$ $\qquad\qquad$ $\cos 15° = \dfrac{1}{\sqrt{2}}.\dfrac{\sqrt{3}}{2}+\dfrac{1}{\sqrt{2}}\dfrac{1}{2}$

$\qquad\qquad\qquad\qquad = \dfrac{\sqrt{3}+1}{2\sqrt{2}}$

$\qquad\qquad\qquad\qquad = 0.966$

This confirms the fact that $\quad \sin 75° = \cos 15°$.

In solving some trigonometrical equations the compound angle formulae are used together with the following trigonometrical identities:

$$\tan x = \frac{\sin x}{\cos x} \qquad \sin^2 x+\cos^2 x = 1$$

EXAMPLE 4

Find the angle between $0°$ and $90°$ which satisfies the

equation $\qquad\qquad$ $\sin x = 2 \sin (x-20°)$

We have, $\qquad\qquad$ $\sin x = 2 \sin (x-20°)$

$\qquad\qquad\qquad = 2\{\sin x.\cos 20°-\cos x.\sin 20°\}$

$\qquad\qquad\qquad = 2\{(\sin x)0.9397-(\cos x)0.3420\}$

$\qquad\qquad\qquad = 1.8794(\sin x)-0.6840(\cos x)$

$\therefore$ $\qquad$ $0.6840 \cos x = 1.8794 \sin x-\sin x$

$\therefore$ $\qquad$ $0.6840 \cos x = 0.8794 \sin x$

$\therefore$ $\qquad\qquad$ $\dfrac{\sin x}{\cos x} = \dfrac{0.6840}{0.8794}$

$\therefore$ $\qquad\qquad$ $\tan x = 0.7778$

$\therefore$ $\qquad\qquad$ $x = 37° 53'$

Exercise 23

1) Simplify:

(a) $\sin (x+180°)$ (b) $\cos (180°-x)$

(c) $\sin (90°-x)$ (d) $\cos (270°+x)$

2) Using the formula for $\sin (A+B)$ find a formula for $\sin 2\theta$.

3) Show that $\sin (\phi+60°)+\sin (\phi-60°) = \sin \phi$.

4) If $\tan \theta = \dfrac{3}{4}$ and $\tan \phi = \dfrac{5}{12}$ sketch suitable right angled triangles from which the other trigonometrical ratios of θ and ϕ may be found. Hence find the values of:

(a) $\sin (\theta-\phi)$ (b) $\cos (\theta+\phi)$

5) Show that $\sqrt{2} \sin \left(\theta-\dfrac{\pi}{4}\right) = \sin \theta-\cos \theta$.

6) Find the angle between $0°$ and $90°$ which satisfies the equation:

$$\cos \theta = 3 \cos (\theta+30°)$$

7) Find the value of sin 15° using the fact $\sin 15° = \sin(45°-30°)$.

8) If $\cos \alpha = \dfrac{3}{5}$ and $\cos \beta = \dfrac{4}{5}$ find the value of $\cos (\alpha+\beta)$.

9) Find the angle between $0°$ and $90°$ which satisfies the equation $2 \cos (\theta+60°) = \sin (\theta+30°)$.

10) If $\sin 33° \ 24' = 0.5505$ find, by using the compound angle formulae, the values of:

(a) $\cos 236° \ 36'$ (b) $\sin 326° \ 36'$.

THE FORM R.sin $(\theta\pm\alpha)$

Consider $3 \sin (\theta+40°)$.

Now $\sin (A+B) = \sin A.\cos B+\cos A.\sin B$

$\therefore$ $3 \sin (\theta+40°) = 3[\sin \theta.\cos 40°+\cos \theta.\sin 40°]$

$$= [3(\sin \theta)0.7660+(\cos \theta)0.6428]$$

$$= 2.298 \sin \theta+1.928 \cos \theta$$

Hence, using the same method, if R and α are constants

then
$$R \sin (\theta \pm \alpha) = R[\sin \theta.\cos \alpha \pm \cos \theta.\sin \alpha]$$
$$= (R.\cos \alpha) \sin \theta \pm (R.\sin \alpha) \cos \theta$$
$$= a.\sin \theta \pm b.\cos \theta$$

where
$$a = R.\cos \alpha \qquad \qquad [1]$$

and
$$b = R.\sin \alpha \qquad \qquad [2]$$

$\therefore$ squaring and adding equations [1] and [2]

then $\quad (R.\cos \alpha)^2 + (R.\sin \alpha)^2 = a^2 + b^2$

$\therefore \qquad R^2\cos^2\alpha + R^2\sin^2\alpha = a^2 + b^2$

$\therefore \qquad R^2[\cos^2\alpha + \sin^2\alpha] = a^2 + b^2$

and since $\quad \cos^2\alpha + \sin^2\alpha = 1$

then $\qquad\qquad R^2 = a^2 + b^2$

Also dividing equation [2] by equation [1]

then
$$\frac{R.\sin \alpha}{R.\cos \alpha} = \frac{b}{a}$$

$\therefore \qquad\qquad \tan \alpha = \frac{b}{a}$

Hence

$$R.\sin (\theta \pm \alpha) = a.\sin \alpha \pm b.\cos \alpha$$

where

$$R^2 = a^2 + b^2 \quad \text{and} \quad \tan \alpha = \frac{b}{a}$$

Problems using the above relationship usually occur in the reverse order to the sequence above.

EXAMPLE 5

Express $\quad 4.\sin \theta - 3.\cos \theta \quad$ in the form $\quad R.\sin (\theta - \alpha)$

Comparing the given expression with $\quad a.\sin \theta - b.\cos \theta$

we have $\quad a = 4 \quad$ and $\quad b = 3$,

and since $\quad R^2 = a^2 + b^2 \qquad\qquad$ and also $\quad \tan \alpha = \frac{b}{a}$

then $\qquad R^2 = 4^2 + 3^2 \qquad\qquad \therefore \qquad \tan \alpha = \frac{3}{4}$

$$\therefore \qquad R = \sqrt{16+9} \qquad\qquad\qquad = 0.75$$

$$\therefore \qquad R = 5 \qquad\qquad \therefore \qquad \alpha = 36°\,52'$$

Hence $\qquad 4.\sin\theta - 3.\cos\theta = 5.\sin(\theta - 36°\,52')$

EXAMPLE 6

Express $\quad 2.\sin 3t + 4.\cos 3t \quad$ in the form $\quad R.\sin(3t+\alpha)$.

This example is similar to Example 5 except that the angles $3t$ and α are in radians.

Comparing the given expression with $\quad a.\sin\theta + b.\cos\theta$

we have $\quad a = 2 \quad$ and $\quad b = 4,$

and since $\quad R^2 = a^2 + b^2 \qquad$ and also $\quad \tan\alpha = \dfrac{b}{a}$

then $\qquad R^2 = 2^2 + 4^2 \qquad \therefore \qquad \tan\alpha = \dfrac{4}{2}$

$$\therefore \qquad R = \sqrt{4+16} \qquad\qquad\qquad = 2$$

$$\therefore \qquad R = 4.472 \qquad \therefore \qquad \alpha = 1.107\text{ rad}$$

Hence $\qquad 2.\sin 3t + 4.\cos 3t = 4.472\sin(3t+1.107)$

EXAMPLE 7

Express $\quad 6\sin\omega t - 8\cos\omega t \quad$ in the form $\quad R.\sin(\theta-\alpha)$. Hence find the maximum value of $\quad 6\sin\omega t - 8\cos\omega t \quad$ and the value of ωt at which it occurs.

We have $\quad a = 6 \quad$ and $\quad b = 8$

Then $\qquad R = \sqrt{6^2+8^2} = 10 \quad$ and $\quad \tan\omega t = \dfrac{8}{6} = 1.333$

$$\omega t = 0.927\text{ rad}$$

Hence $\quad 6\sin\omega t - 8\cos\omega t = 10\sin(\omega t - 0.927).$

Now $\quad 6\sin\omega t - 8\cos\omega t \quad$ will be a maximum

when, $\qquad \sin(\omega t - 0.927) = 1$

i.e. when, $\qquad \omega t - 0.927 = \dfrac{\pi}{2} = 1.571$

or $\qquad\qquad \omega t = 1.571 + 0.927 = 2.498$

Therefore, the maximum value of $\quad 6\sin\omega t - 8\cos\omega t \quad$ is 10 and it occurs when $\quad \omega t = 2.498$ rad.

Exercise 24

1) Express $3.\sin\theta + 2.\cos\theta$ in the form $R.\sin(\theta + \alpha)$.

2) Express $7.\cos\omega t + \sin\omega t$ in the form $R.\sin(\omega t + \alpha)$.

3) Rewrite $5.\sin\omega t - 7.\cos\omega t$ in the form $R.\sin(\omega t - \alpha)$.

4) Using the result obtained in Question (3) find the maximum value of $5.\sin\omega t - 7.\cos\omega t$ and the value of ωt at which this occurs.

5) An electric current is given by $i = 200.\sin 300t + 100.\cos 300t$. Express this as a single trigonometrical function and find its maximum value.

6) If $y = 5\sin(x+30°) + 10\cos(x-30°)$ express y in the forms:
(a) $a\sin x + b\cos x$ (b) $R\sin(x+\alpha)$.

SUMMARY

$$\sin(A+B) = \sin A.\cos B + \cos A.\sin B$$

$$\sin(A-B) = \sin A.\cos B - \cos A.\sin B$$

$$\cos(A+B) = \cos A.\cos B - \sin A.\sin B$$

$$\cos(A-B) = \cos A.\cos B + \sin A.\sin B$$

$$R.\sin(\theta \pm \alpha) = a.\sin\theta \pm b.\cos\theta$$

where $$R^2 = a^2 + b^2$$

and $$\tan\alpha = \frac{b}{a}$$

 # THE BINOMIAL THEOREM

On reaching the end of this chapter you should be able to:-

1. Expand expressions of the form $(a+x)^n$ for small, positive integer n.
2. State the general form for the binomial coefficients for all positive integer n.
3. Solve problems using the binomial theorem relevant to problems in technology, to include approximations.

BINOMIAL EXPRESSION

A **binomial expression** consists of two terms. Thus $1+x$, $a+b$, $5y-2$, $3x^2+7$ and $7a^3+3b^2$ are all binomial expressions. The **binomial theorem** allows us to expand powers of such expressions.

THE BINOMIAL THEOREM

Now $(a+b)^0 = 1$

(since any number to the power 0 is unity),

and also $(a+b)^1 = a+b$

By multiplying, $(a+b)^2 = a^2+2ab+b^2$

also $(a+b)^3 = a^3+3a^2b+3ab^2+b^3$

and $(a+b)^4 = a^4+4a^3b+6a^2b^2+4ab^3+b^4$

We can now arrange the coefficients of each of the terms of the above expansions in the form known as **Pascal's triangle**

Binomial expression	Coefficients in the expansion
$(a+b)^0$	1
$(a+b)^1$	1　1
$(a+b)^2$	1　2　1
$(a+b)^3$	1　3　3　1
$(a+b)^4$	1　4　6　4　1
$(a+b)^5$	1　5　10　10　5　1
$(a+b)^6$	1　6　15　20　15　6　1
$(a+b)^7$	1　7　21　35　35　21　7　1

It will be seen that:

(a) The number of terms in each expansion is one more than the index. Thus the expansion of $(a+b)^9$ will have 10 terms.

(b) The arrangement of the coefficients is symmetrical.

(c) The coefficients of the first and last terms are both always unity.

(d) Each coefficient in the table is obtained by adding together the two coefficients in the line above which lie on either side of it.

The expansion of $(a+b)^8$ is therefore:

$$a^8+(1+7)a^7b+(7+21)a^6b^2+(21+35)a^5b^3+(35+35)a^4b^4$$

$$+(35+21)a^3b^5+(21+7)a^2b^6+(7+1)ab^7+b^8$$

$$= a^8+8a^7b+28a^6b^2+56a^5b^3+70a^4b^4+56a^3b^5+28a^2b^6+8ab^7+b^8$$

It is inconvenient to use Pascal's traingle when expanding higher powers of $(a+b)$. In such cases the following series is used.

$$(a+b)^n = a^n+na^{n-1}b+\frac{n(n-1)}{2!}a^{n-2}b^2+\frac{n(n-1)(n-2)}{3!}a^{n-3}b^3+\ldots+b^n$$

This is the **binomial theorem** and is true for all positive whole numbers n.

The symbol ! indicates 'factorial' when following a positive whole number.

For example, 4! is pronounced 'factorial four' and means $4\times3\times2\times1$. Thus　$2! = 2\times1$,　　$3! = 3\times2\times1$,　　$4! = 4\times3\times2\times1$, $5! = 5\times4\times3\times2\times1$,　and so on.

EXAMPLE 1

Expand　$(3x+2y)^4$.

Comparing　$(3x+2y)^4$　with　$(a+b)^4$,　we have $3x$ in place of a, and $2y$ in place of b. Substituting in the standard expansion, we get:

$$[(3x)+(2y)]^4 = (3x)^4+4(3x)^3(2y)+\frac{(4)(4-1)}{2!}(3x)^2(2y)^2$$

$$+\frac{(4)(4-1)(4-2)}{3!}(3x)(2y)^3+(2y)^4$$

$$= (3x)^4+4(3x)^3 2y+\frac{4\times3}{2\times1}(3x)^2(2y)^2+\frac{4\times3\times2}{3\times2\times1}(3x)(2y)^3+(2y)^4$$

$$= 81x^4+216x^3y+216x^2y^2+96xy^3+16y^4$$

EXAMPLE 2

Find the first four terms of the expansion of $(x-4y)^{15}$.

The given binomial expression should be rewritten as $[x+(-4y)]^{15}$ and comparing this with the standard expression for $(a+b)^n$ we have x in place of a, $-4y$ in place of b, and 15 in place of n.

$$\therefore \quad [x+(-4y)]^{15} = x^{15}+15x^{14}(-4y)+\frac{(15)(15-1)}{2!}x^{13}(-4y)^2$$

$$+\frac{15(15-1)(15-2)}{3!}x^{12}(-4y)^3$$

$$= x^{15}+15(-4)x^{14}y+\frac{15\times14\times(-4)^2}{2\times1}x^{13}y^2$$

$$+\frac{15\times14\times13\times(-4)^3}{3\times2\times1}x^{12}y^3$$

$$= x^{15}-60x^{14}y+1680x^{13}y^2-29\ 120x^{12}y^3$$

APPLICATION TO SMALL ERRORS

In the binomial theorem if we put $a = 1$ and $b = x$ we get

$$(1+x)^n = 1+nx+\frac{n(n-1)}{2!}x^2+\frac{n(n-1)(n-2)}{3!}x^3+ \ldots$$

When x is very small,

$$(1+x)^n = 1+nx \quad \text{approximately,}$$

as all other terms in the series contain powers of x which are negligible when compared with the first two shown. This relationship is often useful.

It is easy enough to measure the expansion of a metal rod which has been heated. It is, however, not so easy to measure the increases in areas and volumes which have been expanded.

If α is the coefficient of linear expansion, then the length of an expanded bar originally 1 m long is $(1+\alpha)$ m   for one degree temperature increase. An area of the same material originally $1 \text{ m} \times 1 \text{ m}$ becomes $(1+\alpha)^2$ m² when it expands. From the above approximation as α is very small then:

$$(1+\alpha)^2 = 1+2\alpha$$

Hence *the area coefficient is approximately twice the linear coefficient.*

Similarly a volume $1 \text{ m} \times 1 \text{ m} \times 1 \text{ m}$ on expanding becomes $(1+\alpha)^3$ m³ and again, using the above approximation:

$$(1+\alpha)^3 = 1+3\alpha$$

which shows that *the volume coefficient is approximately three times the linear coefficient.*

EXAMPLE 3

In measuring the radius of a circle, the measurement is 1% too large. If this measurement is used to calculate the area of the circle find the resulting error.

Let A be the area of the circle, and r the radius. If δA is the error in the area, then:

$$A = \pi r^2$$

and
$$A+\delta A = \pi(r+r \times \tfrac{1}{100})^2$$
$$= \pi r^2(1+\tfrac{1}{100})^2$$

and since $\tfrac{1}{100}$ is small when compared with 1, then:

$$A+\delta A = \pi r^2(1+2 \times \tfrac{1}{100})$$
$$= A(1+\tfrac{2}{100})$$
$$\delta A = \tfrac{2}{100}A$$

Hence
$$\delta A = 2\% \text{ of } A$$

Thus an error of 1% too large in the radius gives an error of 2% too large in the area.

EXAMPLE 4

When a uniform beam is simply supported at its ends the deflection at the centre of span is given by,

$$y = \frac{5wl^4}{384EI}$$

where w is the distributed load per unit length, l is the length between supports, E is Young's modulus and I is the 2nd moment of cross-sectional area. Find the percentage change in y when l decreases by 3%.

Let the change in y be δy. Then:

$$y+\delta y = \frac{5w}{384EI}[l-l\times\tfrac{3}{100}]^4$$

$$= \frac{5wl^4}{384EI}[1+(-\tfrac{3}{100})]^4$$

$$= \frac{5wl^4}{384EI}[1+4(-\tfrac{3}{100})]$$

and approximating since $\tfrac{3}{100}$ is small compared with 1:

$$y+\delta y = y[1-\tfrac{12}{100}]$$

$$\therefore \qquad \delta y = -\tfrac{12}{100}y \quad \text{or} \quad 12\% \text{ of } y \text{ (decrease)}$$

Hence y will decrease by 12% approximately.

EXAMPLE 5

A formula used in connection with close-coiled helical springs is:

$$P = \frac{GFd^5}{8hD^3}$$

Find the change in P if d is increased by 2% and D is decreased by 3%.
Let δP be the change in P. Then:

$$P+\delta P = \frac{GF(d+\tfrac{2}{100}d)^5}{8h(D-\tfrac{3}{100}D)^3}$$

$$= \frac{GFd^5(1+\tfrac{2}{100})^5}{8hD^3(1-\tfrac{3}{100})^3}$$

$$= P\times\frac{(1+5\times\tfrac{2}{100})}{(1-3\times\tfrac{3}{100})}$$

since both $\tfrac{2}{100}$ and $\tfrac{3}{100}$ are small when compared with 1:

$$= P\times\frac{(1+\tfrac{10}{100})}{(1-\tfrac{9}{100})}$$

$$= P\times\tfrac{110}{91}$$

$$= P\times(1.21)$$

$$= P\times(1+0.21)$$

$$= P+\tfrac{21}{100}P$$

$$\therefore \qquad \delta P = 21\% \text{ of } P$$

Hence P will increase by 21% approximately.

Exercise 25

Expand using Pascal's triangle:

1) $(1+z)^5$

2) $(p+q)^6$

3) $(x-3y)^4$

4) $(2p-q)^5$

5) $(2x+y)^7$

6) $\left(x+\dfrac{1}{x}\right)^3$

Find the first four terms of the following expansions using the binomial theorem formula:

7) $(1+x)^{12}$

8) $(1-2x)^{14}$

9) $(p+q)^{16}$

10) $(1+3y)^{10}$

11) $(x^2-3y)^9$

12) $\left(x^2+\dfrac{1}{x^2}\right)^{11}$

13) Show that an error of 1% in the measurement of the radius of a sphere leads to an error of approximately 2% in the outer surface area, and 3% in the volume.

14) Find the approximate percentage error in the calculated volume of a right circular cone if the radius is taken as 2% too large, and the height is taken as 3% too small.

15) A formula used in connection with helical springs is $y = \dfrac{8WnD^3}{Gd^4}$.

Find the percentage error in y if D is 1% too small, and d is 2% too large.

16) In the standard gas equation $\dfrac{pv}{T} = k$, the volume v is increased by 2%, the temperature T is diminished by 1%, and the value of constant k remains unaltered. What is the corresponding percentage change in the pressure p?

17) The deflection y at the centre of a steel rod of length l and circular cross-section of diameter d, simply supported at its ends and carrying a concentrated load W at its centre is given by the formula $y = \dfrac{Wl^3}{d^4}$

Find the percentage change in y when l increases by 2% and d decreases by 3%.

SUMMARY

The general expression of the **Binomial Theorem** for all positive whole numbers n:

$$(a+b)^n = a^n + na^{n-1}b + \frac{n(n-1)}{2\times1}a^{n-2}b^2 + \frac{n(n-1)(n-2)}{3\times2\times1}a^{n-3}b^3 + \ldots + b^n$$

13. THE EXPONENTIAL FUNCTION

On reaching the end of this chapter you should be able to:-

1. *Define the exponential function in terms of its differential property.*
2. *State the expansion of the exponential function in a power series.*
3. *Plot the graphs of the functions* $y = e^x$ *and* $y = e^{-x}$, *using tables.*

4. *Solve problems in growth and decay arising from technology, to include charge and discharge of a capacitor, growth and decay of a current in an inductor, and radioactive decay.*
5. *Define Naperian logarithms.*
6. *Determine Naperian logarithms from tables.*

THE EXPONENTIAL FUNCTION (e^x)

e^x is called 'the exponential function'. It may be expressed as an infinite series, that is:

$$e^x = 1 + x + \frac{x^2}{2!} + \frac{x^3}{3!} + \frac{x^4}{4!} + \ldots$$

Now, if we differentiate this series term by term we obtain:

$$\frac{d}{dx}(e^x) = 1 + \frac{2x}{2!} + \frac{3x^2}{3!} + \frac{4x^3}{4!} + \ldots$$

and considering a typical coefficient, for example $\dfrac{4}{4!} = \dfrac{4}{4 \times 3 \times 2 \times 1} = \dfrac{1}{3!}$

then

$$\frac{d}{dx}(e^x) = 1 + x + \frac{x^2}{2!} + \frac{x^3}{3!} + \ldots$$

which is the same as the original series.

Hence $\dfrac{d}{dx}(e^x) = e^x$ which confirms the result already obtained by a different method on page 12.

This is the only mathematical function which does *not* change on differentiation.

Put another way we may say that for the graph of $y = e^x$ the rate of change $\dfrac{dy}{dx}$, at any point, is equal to e^x, that is $\dfrac{dy}{dx} = y$.

THE NUMERICAL VALUE OF e

Although we have defined the exponential function as e^x we shall not be able to make use of it unless we know the value of the constant e.

This may be found by substituting 1 for x in the series for e^x.

Hence $e = 1 + 1 + \dfrac{1^2}{2!} + \dfrac{1^3}{3!} + \dfrac{1^4}{4!} + \dfrac{1^5}{5!} + \dfrac{1^6}{6!} + \dfrac{1^7}{7!} + \ldots$

$\qquad = 1 + 1 + 0.5 + 0.1667 + 0.0417 + 0.0083 + 0.0014 + 0.0002 + \ldots$

$\qquad = 2.7183$

$\therefore \qquad e = 2.718$ correct to three decimal places.

If greater accuracy is required more terms will have to be considered.

THEORY OF LOGARITHMS

If N is a number such that

$$N = a^x$$

we say that x is the logarithm of N to the base a. We write:

$$\log_a N = x$$

It should be carefully noted that:

$$\text{NUMBER} = \text{BASE}^{\text{LOGARITHM}}$$

Since $125 = 5^3$ we may write $\log_5 125 = 3$
and because $16 = 2^4$ we may write $\log_4 16 = 2$.

EXAMPLE 1

If $\log_7 49 = x$,   find the value of x.

Writing the equation in index form we have,

$$49 = 7^x$$

or

$$7^2 = 7^x$$

Hence $x = 2$ (since the indices on each side of the equation must be the same).

EXAMPLE 2

If $\log_x 8 = 3$, find the value of x.

Writing the equation in index form we have,

$$8 = x^3$$

or $$2^3 = x^3$$

Hence $x = 2$ (since the indices on both sides of the equation are the same, the bases must be the same).

Exercise 26

In each of the following find the value of x:

1) $\log_x 9 = 2$ 5) $\log_3 x = 2$ 9) $\log_x 8 = 3$

2) $\log_x 81 = 4$ 6) $\log_4 x = 3$ 10) $\log_x 27 = 3$

3) $\log_2 16 = x$ 7) $\log_{10} x = 2$

4) $\log_5 125 = x$ 8) $\log_7 x = 0$

LOGARITHMS TO THE BASE 10

Logarithms to the base 10 are called common logarithms. The logarithm to the base 10 of x is written $\log_{10} x$. When using logarithmic tables to solve numerical problems, the values given in the tables will to be the base 10. It is assumed that the reader is familiar with the use of common logarithms.

LOGARITHMS TO THE BASE 'e'

In higher mathematics all logarithms are taken to the base e, where $e = 2.718\ 28$. Logarithms to this base are often called natural logarithms. They are also called Naperian or hyperbolic logarithms.

To avoid confusion the base of a logarithm should always be stated, e.g., $\log_{10} x$, but in practice this is often omitted if it is reasonably obvious which base is being used.

In the examples which follow the base will always be given.

Common logarithms are given as $\log_{10}$ (or lg)

and natural logarithms are given as $\log_e$ (or ln)

CHOICE OF BASE

When using logarithmic tables for numerical calculations common logarithms, i.e., $\log_{10}$ are preferred. This is because they are simpler to use, in table form, than natural logarithms.

If an electronic calculating machine of the scientific type (i.e., having keys which give trignometrical and logarithmic functions in addition to the usual $+$, $-$, $\times$ and $\div$ etc.) is used then it is just as easy to use logarithms to the base e. Some machines have keys for both $\log_e$ and $\log_{10}$ but on the more limited models only $\log_e$ is given.

The natural logarithm is found by using the $\log_e$ (or ln) key and the natural antilogarithm is found by using the e^x key.

NATURAL LOGARITHMIC TABLES

In most books of mathematical tables there is a table of natural logarithms. Part of such a table is shown below:

Hyperbolic, Natural or Naperian Logarithms

4.5	1.5041	5063	5085	5107	5129	5151	5173	5195	5217	5239	2 4 7	9 11 13	15 18 20
4.6	1.5261	5282	5304	5326	5347	5369	5390	5412	5433	5454	2 4 6	9 11 13	15 17 19
4.7	1.5476	5497	5518	5539	5560	5581	5602	5623	5644	5665	2 4 6	8 11 13	15 17 19
4.8	1.5686	5707	5728	5748	5769	5790	5810	5831	5851	5872	2 4 6	8 10 12	14 16 19
4.9	1.5892	5913	5933	5953	5974	5994	6014	6034	6054	6074	2 4 6	8 10 12	14 16 18
5.0	1.6094	6114	6134	6154	6174	6194	6214	6233	6253	6273	2 4 6	8 10 12	14 16 18
5.1	1.6292	6312	6332	6351	6371	6390	6409	6429	6448	6467	2 4 6	8 10 12	14 16 18
5.2	1.6487	6506	6525	6544	6563	6582	6601	6620	6639	6658	2 4 6	8 10 11	13 15 17
5.3	1.6677	6696	6714	6734	6752	6771	6790	6808	6827	6845	2 4 6	7 9 11	13 15 17
5.4	1.6864	6882	6901	6919	6938	6956	6974	6993	7011	7029	2 4 5	7 9 11	13 15 17

Natural logarithms of 10^{+n}

n	1	2	3	4	5	6	7	8	9
$\log_e 10^n$	2.3026	4.6052	6.9078	9.2103	11.5129	13.8155	16.1181	18.4207	20.7233

The first column of a set of full tables gives the natural logarithms of numbers from 1.0 to 9.9 (the specimen table above only gives numbers from 4.5 to 5.49), and the tables are read in the same way as ordinary log tables except that the characteristic is also given. Thus:

$$\log_e 4.568 = 1.519\ 1$$

When a natural logarithm of a number, which lies outside the tabulated range, is required the subsidiary table has to be used. The following examples show how this is done.

EXAMPLE 3

To find $\log_e 483.4$.

$$483.4 = 4.834 \times 100 = 4.834 \times 10^2$$

$\therefore$ $\log_e 483.4 = \log_e 4.834 + \log_e 10^2$

From the main table $\log_e 4.834 = 1.5756$

From the subsidiary table $\log_e 10^2 = 4.6052$

$\therefore$ $\log_e 483.4 = 1.5756 + 4.6052 = 6.1808$

EXAMPLE 4

To find $\log_e 0.053\ 61$.

$$0.053\ 61 = \frac{5.361}{100}$$

$\therefore$ $\log_e 0.053\ 61 = \log_e \left(\frac{5.361}{10^2} \right)$

$\log_e 5.361 = \log_e 10^2$

From the main table $\log_e 5.361 = 1.679\ 2$

From the subsidiary table $\log_e 10^2 = 4.6052$

$\log_e 0.053\ 61 = 1.6792 - 4.6052$

$$= -2.9260$$

VALUES OF e^x AND e^{-x}

Most mathematical tables which include logarithms and trignometrical functions also include a table giving values of e^x and e^{-x} for values of x from 0 to 6. It may appear that the range of x values is rather limited but it is adequate for most practical problems as may be seen from the examples which follow later in this chapter.

EXPONENTIAL GRAPHS

Curves which have equations of the type e^x and e^{-x} are called exponential graphs.

We may plot the graphs of e^x and e^{-x} by using mathematical tables to find values of e^x and e^{-x} for chosen values of x. We should remember that any number to a zero power is unity: hence $e^0 = 1$.

Drawing up a table of values we have:

x	-2	-1	0	1	2
e^x	0.14	0.37	1	2.72	7.39
$-x$	2	1	0	-1	-2
e^{-x}	7.39	2.72	1	0.37	0.14

For convenience both the curves are shown plotted on the same axes in Fig. 13.1. Although the range of values chosen for x is limited, the overall shape of the curves is clearly shown.

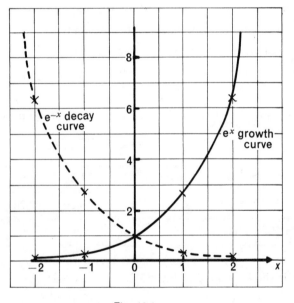

Fig. 13.1

The rate at which a curve is changing at any point is given by the gradient of the tangent at that point.

Remember the sign convention for gradients is:

positive   negative
gradient   gradient

Now the gradient at any point on the e^x graph is positive and so the rate of change is positive. In addition the rate of change increases as the values of x increase. A graph of this type is called a growth curve.

The gradient at any point on the e^{-x} graph is negative and so the rate of change is negative. In addition the rate of change decreases as the values of x decrease. A graph of this type is called a decay curve.

EXAMPLE 5

The instantaneous e.m.f. in an inductive circuit is given by the expression $100.e^{-4t}$ volts, where t is time in seconds. Plot the graph of the e.m.f. for values of t from 0 to 0.5 seconds, and use the graph to find:

(a) the value of the e.m.f. when $t = 0.25$ seconds, and

(b) the rate of change of the e.m.f. when $t = 0.1$ seconds.

Check this result using the method of differentiation.

First we will draw up a table of values taking values of t at 0.1 s intervals:

t	0	0.1	0.2	0.3	0.4	0.5
$-4t$	0	-0.4	-0.8	-1.2	-1.6	-2.0
e^{-4t}	1	0.67	0.45	0.30	0.20	0.14
$100e^{-4t}$	100	67	45	30	20	14

The graph is shown plotted in Fig. 13.2.

(a) The point P on the curve is at 0.25 seconds shown on the t scale and the corresponding value of e.m.f. can be read directly from the vertical axis scale. The value is 37 volts.

(b) The point Q on the graph is at 0.1 seconds. Now the rate of change of the curve at Q is given by the gradient of the tangent at Q. This gradient may be found by constructing a suitable right angled triangle such as MNO in Fig. 13.2, and finding the ratio $\dfrac{MO}{ON}$.

Hence the gradient at Q

$$= \frac{MO}{ON} = \frac{92 \text{ volts}}{0.34 \text{ seconds}} = 270 \text{ volts per second.}$$

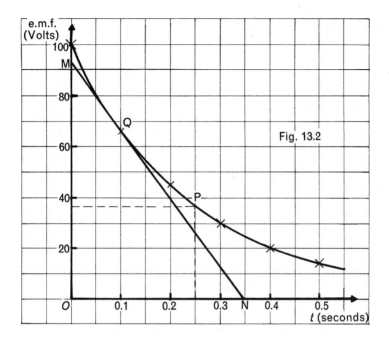

Fig. 13.2

According to the sign convention a line sloping downwards from left to right has a negative gradient.

Hence the gradient at Q is -270 volts per second, which means that the rate of change of the curve at Q is -270 volts per second.

This is the same as saying that the e.m.f. at $t = 0.1$ seconds is decreasing at the rate of 270 volts per second.

A Check Using Differentiation

The rate of change of the e.m.f. at any instant is given by the gradient of the e.m.f.—time curve and using the calculus this is denoted by $\frac{d}{dt}(\text{e.m.f.})$.

Since we have

$$\text{e.m.f.} = 100.e^{-4t}$$

then

$$\frac{d}{dt}(\text{e.m.f.}) = 100(-4)e^{-4t}$$

$$= -400e^{-4t}$$

Hence when $t = 0.1$ seconds

then the required rate of change $= -400 . e^{-4(0.1)}$

$$= -400 . e^{-0.4}$$

$$\doteqdot -268 \text{ volts/second}$$

This verifies the approximate value of -270 volts/second obtained by use of the graph.

EXAMPLE 6

The formula $i = 2(1 - e^{-10t})$ gives the relationship between the instantaneous current i amperes and the time t seconds in an inductive circuit. Plot a graph of i against t, taking values of t from 0 to 0.3 seconds at intervals of 0.05 seconds. Hence find:

(a)   the initial rate of growth of the current i when $t = 0$,   and

(b)   the time taken for the current to increase from 1 to 1.6 amperes.

Verify these results using theoretical methods.

The table of values is drawn up as follows:

t	0	0.05	0.10	0.15	0.20	0.25	0.30
$-10t$	0	-0.5	-1.0	-1.5	-2.0	-2.5	-3.0
1	1	1	1	1	1	1	1
e^{-10t}	1	0.61	0.37	0.22	0.14	0.08	0.05
$(1 - e^{-10t})$	0	0.39	0.63	0.78	0.86	0.92	0.95
$2(1 - e^{-10t})$	0	0.78	1.26	1.56	1.72	1.84	1.90

The curve is shown plotted in Fig. 13.3.

(a)   When $t = 0$ the initial rate of growth will be given by the gradient of the tangent at O. The tangent at O is the line OM and its gradient may be found by using a suitable right angled triangle MNO and finding the ratio $\dfrac{\text{MN}}{\text{ON}}$.

Hence the initial rate of growth of $i = \dfrac{\text{MN}}{\text{ON}} = \dfrac{2 \text{ amperes}}{0.1 \text{ seconds}}$

$$= 20 \text{ amperes per second.}$$

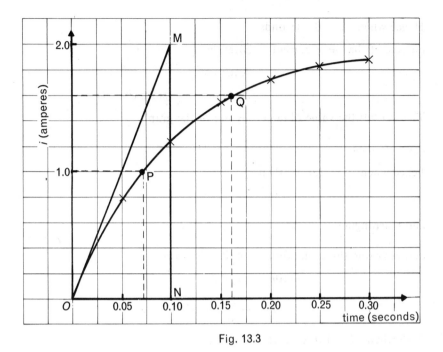

Fig. 13.3

(b) The point P on the curve corresponds to a current of 1.0 amperes and the time at which this occurs may be read from the t scale and is 0.07 seconds.

Similarly point Q corresponds to a current of 1.6 ampere and occurs at 0.16 seconds.

Hence the time between P and Q is $0.16 - 0.07 = 0.09$ seconds.

This means that the time for the current to increase from 1 to 1.6 amperes is 0.09 seconds.

Solution using Theoretical Methods

(a) The rate of growth of current is another way of stating a positive rate of change of current. At any instant this is given by the gradient of the current-time curve and using the calculus this is denoted by $\dfrac{d}{dt}$ (current).

Since we have   current $= 2(1 - e^{-10t})$

or $i = 2 - 2.e^{-10t}$

then $\dfrac{di}{dt} = -2(-10)e^{-10t} = 20e^{-10t}$

Hence when $t = 0$ seconds

the rate of change of current $= 20.e^{-10(0)}$

$$= 20.e^0$$

$$= 20 \text{ A/s} \qquad \text{(since } e^0 = 1)$$

This verifies the value obtained by use of the graph.

(b) We have $\qquad i = 2(1-e^{-10t})$

hence when $i = 1.6$ A then

$$1.6 = 2(1-e^{-10t})$$

and rearranging $\qquad e^{-10t} = 1-\dfrac{1.6}{2} = 0.2$

and rewriting this equation in logarithmic form

$$-10t = \log_e 0.2$$

$\therefore \qquad\qquad t = -\dfrac{1}{10}\log_e 0.2 = 0.161 \text{ s}$

Also when $i = 1.0$ A then

$$1 = 2(1-e^{-10t})$$

and rearranging $\qquad e^{-10t} = 1-\dfrac{1}{2} = 0.5$

and rewriting this equation in logarithmic form

$$-10t = \log_e 0.5$$

$\therefore \qquad\qquad t = -\dfrac{1}{10}\log_e 0.5 = 0.069 \text{ s}$

Now

required time $=$ (time to increase to 1.6 A)$-$(time to increase to 1.0 A)

$$= 0.161-0.069 = 0.092 \text{ s}$$

This verifies the more approximate value obtained by use of the graph.

Exercise 27

1) Find the values of:

(a) $\log_e 9.137$ (b) $\log_e 6.017$

(c) $\log_e 29.16$ (d) $\log_e 298.2$

(e) $\log_e 0.5917$ (f) $\log_e 0.007\ 138$

2) A formula used for calculating the self-inductance of parallel conductors is:

$$L = 0.000\ 644\left(\log_e\frac{d}{r}+0.25\right)$$

Find L when $d = 50$ and $r = 0.45$.

3) The formula:

$$L = 0.002\ \log_e\frac{4l}{d}-0.002$$

is used in connection with straight aerials. Calculate the value of L when $l = 250$ and $d = 0.25$.

4) Find the values of:

(a) $e^{0.3}$ (b) $e^{2.5}$ (c) $e^{4.5}$

(d) $e^{-0.4}$ (e) e^{-1} (f) $e^{-3.5}$

5) Plot a graph of $y = e^{2x}$ for values of x from -1 to $+1$ at 0.25 unit intervals. Use the graph to find the value of y when $x = 0.3$, and the value of x when $y = 5.4$.

6) Using values of x from -4 to $+4$ at one unit intervals plot a graph of $y = e^{-x/2}$. Hence find the value of x when $y = 2$, and the gradient of the curve when $x = 0$.

7) For a constant pressure process on a certain gas the formula connecting the absolute temperature T and the specific entropy s is $T = 24.e^{3s}$. Plot a graph of T against s taking values of s equal to 1.000, 1.033, 1.066, 1.100, 1.133, 1.166 and 1.200. Use the graph to find the value of:

(a) T when $s = 1.09$.

(b) s when $T = 700$.

8) The equation $i = 2.4e^{-6t}$ gives the relationship between the instantaneous current, i mA, and the time, t seconds. Plot a graph of i against t for values of t from 0 to 0.6 seconds at 0.1 second intervals. Use the curve obtained to find the rate at which the current is decreasing when $t = 0.2$ seconds. (The more able student may wish to check the results obtained graphically by a theoretical method using calculus.)

9) In a capacitive circuit the voltage v and the time t seconds are connected by the relationship $v = 240(1-e^{-5t})$. Draw the curve of v against t for values of $t = 0$ to $t = 0.7$ seconds at 0.1 second intervals.

Hence find:

(a) the time when the voltage is 140 volts, and

(b)   the initial rate of growth of the voltage when $t = 0$.

(The more able student may wish to check the results obtained graphically by a theoretical method using calculus.)

10)   A   radioactive   material   decays   according to   the   law $N = N_i(e^{-0.0693t})$ where N is   the   amount of radioactivity remaining, N_i is the initial amount of radioactivity, and t years is the time taken to decay from N_i to N. Plot a graph of N (this will be in terms of N_i, e.g., $0.4N_i$) against t for values of $t = 0$ to $t = 30$ years   at 5 yearly intervals. Hence find the half life of the material (that is when half the radioactivity has decayed or $N = \frac{1}{2}N_i$).

SUMMARY

The exponential series is:

$$e^x = 1 + x + \frac{x^2}{2!} + \frac{x^3}{3!} + \frac{x^4}{4!} + \dots$$

A property of e^x is that it remains unaltered by differentiation that is:

$$\frac{d}{dx}(e^x) = e^x$$

The numerical value of e is 2.718 (correct to three decimal places).

$$\text{NUMBER} = \text{BASE}^{\text{LOGARITHM}}$$

hence if $N = a^x$

then $\log_a N = x$

Common logarithms are given as $\log_{10}$ or   lg

Natural logarithms are given as $\log_e$ or   ln

LOGARITHMIC GRAPH PAPER

On reaching the end of this chapter you should be able to :-

1. *Plot experimental data on log–log and log–linear graph paper.*

2. *Determine laws of experimental data of the form* $z = at^n$, $z = a.b^t$, *and* $z = a.e^{bt}$.

NON-LINEAR LAWS WHICH CAN BE REDUCED TO THE LINEAR FORM

Many non-linear equations can be reduced to the linear (i.e., straight line) form by making a suitable substitution. We have met previously equations such as $y = m.t^2 + c$ and we put $x = t^2$ so that the equation becomes $y = mx + c$ which is a straight line when plotted and enables us to find values of the constants m and c.

This section is similar except that logarithmic and exponential relationship are covered.

THE STRAIGHT LINE

You will remember the standard equation of the straight line is $y = m.x + c$ where m and c are constants.

The values of these constants may be found using information obtained from either Fig. 14.1 or Fig. 14.2.

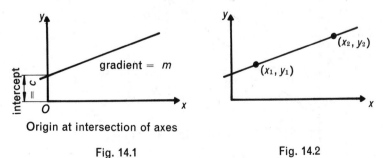

Origin at intersection of axes

Fig. 14.1   Fig. 14.2

171

If the origin is at the intersection of the axes the values of m and c may be found using Fig. 14.1.

If the origin is not shown the two point method must be used (Fig. 14.2), i.e., since the point (x_1, y_1) lies on the line then $y_1 = mx_1 + c$ and the point (x_2, y_2) lies on the line then $y_2 = mx_2 + c$.

These equations may be solved simultaneously to find the values of m and c.

The two point method may be used whether or not the origin is shown and should be used in all the work which follows in this chapter.

EXAMPLE 1

The table below gives corresponding values of x and y. Plot these values on a suitable graph and hence find the equation connecting x and y.

x	3	5	8	12
y	15	23	35	51

The graph is shown plotted in Fig. 14.3.

Since the plotted points lie on a straight line the equation connecting x and y will be of the form $y = mx + c$.

To find the values of the constants m and c choose two points P and Q which lie on the line. The co-ordinates of P and Q will satisfy $y = mx + c$.

Hence at the point Q (10, 43) then $\qquad$ $43 = m(10) + c$ $\qquad$ [1]

and at the point P (4, 19) then $\qquad$ $19 = m(4) + c$ $\qquad$ [2]

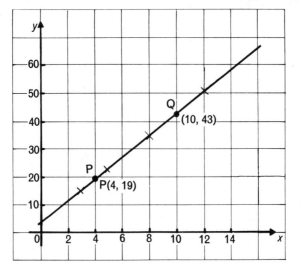

Fig. 14.3

Subtracting equation [2] from [1], $24 = 6m$

$\therefore$ $m = 4$

and substituting $m = 4$ into equation [1]

then $43 = 4(10) + c$

$\therefore$ $c = 3$

Hence the required equation is $y = 4x + 3.$

LOGARITHMIC SCALES

It has been shown earlier that: Number $= $ (base)^{logarthim}

and if we use a base of 10 then: Number $= $ (10)^{logarithm}

Since $1000 = 10^3$ then we may write $\log_{10}1000 = 3$

and since $100 = 10^2$ then we may write $\log_{10}100 = 2$

and since $10 = 10^1$ then we may write $\log_{10}10 = 1$

and since $1 = 10^0$ then we may write $\log_{10}1 = 0$

and since $0.1 = 10^{-1}$ then we may write $\log_{10}0.1 = -1$

and since $0.01 = 10^{-2}$ then we may write $\log_{10}0.01 = -2$

and since $0.001 = 10^{-3}$ then we may write $\log_{10}0.001 = -3$ and so on.

These logarithms may be shown on a scale as shown in Fig. 14.4.

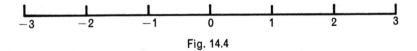

Fig. 14.4

However since we wish to plot numbers directly on to the scale (without any reference to their logarithms) the scale is labelled as shown in Fig. 14.5.

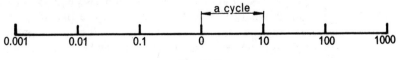

Fig. 14.5

Each division is called a cycle and is sub-divided using a logarithmic scale as, for instance, the scales on a slide rule. Two such cycles are shown in Fig. 14.6.

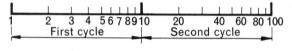

Fig. 14.6

The choice of numbers on the scale depends on the numbers allocated to the variable in the problem to be solved. Thus in Fig. 14.6 the numbers run from 1 to 100.

In all the work which follows in this chapter the logarithms used will be to the base 10. The 10 is often omitted and so, for instance, $\log x$ means $\log_{10} x$.

Consider the following relationships in which z and t are the variables, whilst a, b, and n are constants.

$z = a.t^n$

Now $\qquad\qquad z = a.t^n$

and taking logs $\qquad \log z = \log (a.t^n)$

$$= \log t^n + \log a$$

$\therefore \qquad\qquad \log z = n.\log t + \log a$

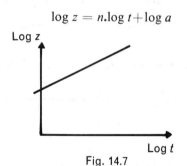

Fig. 14.7

The given values of the variables will satisfy this equation if they satisfy the original equation. Comparing this equation with $y = mx + c$, which is the standard equation of a straight line, we see that if we plot $\log z$ on the y-axis and $\log t$ on the x-axis the result will be a straight line (Fig. 14.7) and the values of the constants n and a may be found using the two point method.

$z = a.b^t$

Now $$z = a.b^t$$

and taking logs $$\log z = \log (a.b^t)$$

$$= \log b^t + \log a$$

$\therefore$ $$\log z = (\log b)t + \log a$$

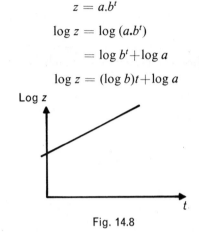

Fig. 14.8

We now proceed in a manner similar to that used for the previous equation by plotting $\log z$ on the y-axis and t on the x-axis, and again obtain a straight line (Fig. 14.8).

$z = ae^{.bt}$

Now $$z = a.e^{bt}$$

and taking logs $$\log z = \log (a.e^{bt})$$

$$= \log e^{bt} + \log a$$

$$= (b.\log e)t + \log a$$

but $\log e = 0.4343$

$\therefore$ $$\log z = (0.4343b)t + \log a$$

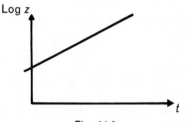

Fig. 14.9

Again proceeding in a manner similar to that used for the previous equations, we plot $\log z$ on the y-axis and t on the x-axis and again obtain a straight line (Fig. 14.9).

LOGARITHMIC GRAPH PAPER

Logarithmic scales (such as those used on a slide rule) may be used on graph paper in place of the more usual linear scales. By using graph paper ruled in this way log plots may be made without the necessity of looking up the logs of each given value. Semi-logarithmic graph paper is also available and has one way ruled with log scales whilst the other way has the usual linear scale. Examples of each are shown in Figs. 14.10 and 14.11.

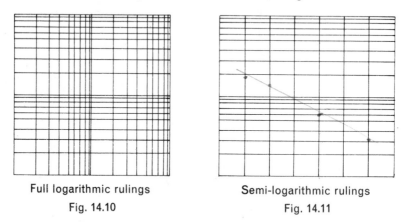

Full logarithmic rulings

Fig. 14.10

Semi-logarithmic rulings

Fig. 14.11

The use of logarithmic graph paper enables power or exponential relationships between two variables to be verified quickly.

A straight line graph on full logarithmic (or log–log) graph paper indicates a relationship between the variables z and t of the form $z = a.t^n$, a and n being constants.

A straight line graph on semi-logarithmic (or log–linear) graph paper indicates a relationship between the variables z and t of the form $z = a.b^t$ or $z = a.e^{bt}$, a and b being constants.

The use of these scales and the special graph paper is shown by the examples which follow.

EXAMPLE 2

The law connecting two quantities z and t is of the form $z = a.t^n$. Find the law given the following pairs of values:

z	3.170	4.603	7.499	10.50	15.17
t	7.980	9.863	13.03	15.81	19.50

The relationship $z = a.t^n$

gives (see text) $\log z = n.\log t + \log a$ [1]

Hence we plot the given values of z and t on log scales as shown in Fig. 14.12

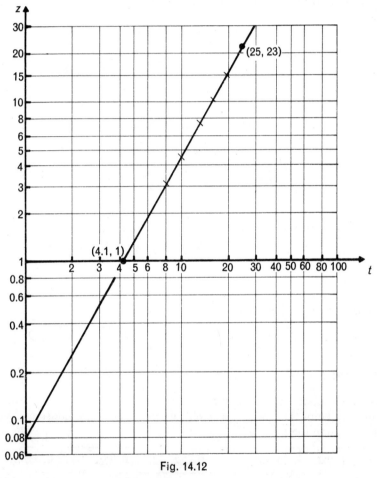

Fig. 14.12

On both the vertical and horizontal axes we require 2 cycles, the first for values from 1 to 10 and the second for values 10 to 100.

The constants are found by taking two pairs of coordinates:

Point $(25, 23)$ lies on the line and putting these values in equation [1],

$$\log 23 = n\{\log 25\} + \log a \qquad [2]$$

Point (4.1, 1) lies on the line and putting these values in equation [1],

$$\log 1 = n\{\log 4.1\} + \log a \qquad [3]$$

Subtracting equation [3] from equation [2],

$$\log 23 - \log 1 = n\{\log 25 - \log 4.1\}$$

$$\log (23/1) = n\{\log (25/4.1)\}$$

$$n = \frac{\log (23/1)}{\log (25/4.1)} = \frac{\log 23}{\log 6.1} = \frac{1.367}{0.7853}$$

$$n = 1.74$$

Substituting this value of n in equation [3],

$$\log 1 = 1.74\{\log 4.1\} + \log a$$

∴ $$\log a = \log 1 - 1.74\{\log 4.1\}$$

$$= 0 - 1.075$$

$$= -1.075 = -1 - 1 + 1 - 0.075$$

$$= -2 + 0.925 = \bar{2}.925$$

∴ $$a = 0.084$$

Hence the law is $z = 0.084t^{1.74}$.

EXAMPLE 3

The table gives values obtained in an experiment. It is thought that the law may be of the form $z = a.b^t$,   where a and b are constants. Verify this and find the law.

t	0.190	0.250	0.300	0.400
z	11 220	18 620	26 920	61 660

We think that the relationship is of the form:

$$z = a.b^t$$

which gives (see text),

$$\log z = (\log b)t + \log a \qquad [1]$$

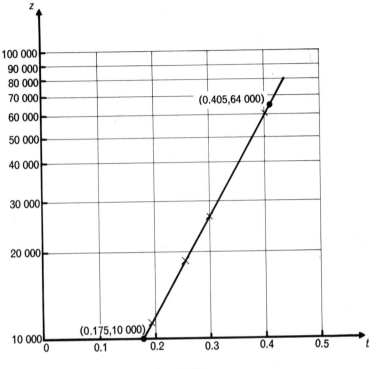

Fig. 14.13

Hence we plot the given values of z on a vertical log scale — the t values however will be on the horizontal axis on an ordinary linear scale (Fig. 14.13).

The points lie on a straight line, and as before the given values of z and t obey the law. We now have to find the coordinates of two points lying on the line.

Point (0.405, 64 000) lies on the line, and substituting in equation [1],

$$\log 64\,000 = (\log b)0.405 + \log a \qquad [2]$$

Point (0.175, 10 000) lies on the line, and substituting in equation [1],

$$\log 10\,000 = (\log b)0.175 + \log a \qquad [3]$$

Subtracting equation [3] from equation [2],

$$\log 64\,000 - \log 10\,000 = (\log b)(0.405 - 0.175)$$

$\therefore$
$$4.8062 - 4.0000 = 0.230\,(\log b)$$

$$\therefore \qquad \log b = \frac{0.8062}{0.230} = 3.5052$$

$$\therefore \qquad b = 3200$$

Substituting in the equation [3],

$$\log 10\,000 = (3.5052)0.175 + \log a$$

$$\therefore \qquad \log a = \log 10\,000 - 3.5(0.175)$$

$$= 4 - 0.6134 = 3.3866$$

$$\therefore \qquad a = 2435$$

Hence the law is:

$$z = 2\,435(3200)^t$$

EXAMPLE 4

V and t are connected by the law $V = a.e^{bt}$. If the values given in the table satisfy the law, find the constants a and b.

t	0.05	0.95	2.05	2.95
V	20.70	24.49	30.27	36.06

The law is:

$$V = a.e^{bt}$$

which gives (see text),

$$\log V = 0.4343bt + \log a \qquad [1]$$

As in last example V values are plotted on a log scale on the vertical axis, whilst the t values are plotted on the horizontal axis on an ordinary linear scale (Fig. 14.14).

Point (3.15, 37.2) lies on the line, and substituting in equation [1],

$$\log 37.2 = (0.4343)b(3.15) + \log a \qquad [2]$$

Point (0.30, 21.5) lies on the line, and substituting in equation [1],

$$\log 21.5 = (0.4343)b(0.3) + \log a \qquad [3]$$

Subtracting,

$$\log 37.2 - \log 21.5 = (0.4343)b(3.15 - 0.3)$$

$$\therefore \qquad b = \frac{\log (37.2/21.5)}{(0.4343)(2.85)} = \frac{0.238}{(0.4343)(2.85)}$$

$$= 0.192$$

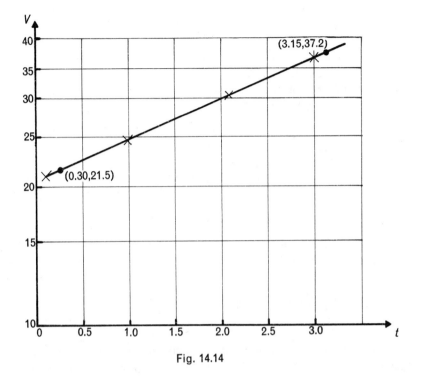

Fig. 14.14

Substituting in the equation [3]

$$\log 21.5 = (0.4343)(0.192)(0.3) + \log a$$

$$\therefore \qquad \log a = 1.3324 - 0.0248 = 1.3076$$

$$\therefore \qquad a = 20.3$$

Exercise 28

1)   The resistance R of the field coils of a motor is measured at different temperatures and found to be as follows:

t (°C)	20	30	40	50	60	70	80
R (ohm)	101	104.6	108.1	111.6	115.1	118.5	122

Plot the graph of R against t. If R and t are related by a law of the form $R = a + bt$,   find from the graph the values of the constants a and b.

2)   In a test on a Lechlanché cell the terminal voltage (V) was measured for various values of current (I) drawn from the cell. The results obtained are shown below.

V	1.44	1.32	1.22	1.06	0.96	0.84
I	0.1	0.3	0.5	0.7	0.9	1.1

Show that the law connecting V and I is of the form $V = E - Ir$ and find values for the constants E (the e.m.f. of the cell) and r (the internal resistance of the cell).

3) Using log–log graph paper show that the following set of values for x and y follows a law of the type $y = ax^n$. From the graph determine the values of the constants a and n.

x	4	16	25	64	144	296
y	6	12	15	24	36	52

4) The following results were obtained in an experiment to find the relationship between the luminosity I of a metal filament lamp and the voltage V:

V	60	80	100	120	140
I	11	20.5	89	186	319

Allowing for the fact than an error was made in one of the readings show that the law between I and V is of the form $I = aV^n$ and find the probable correct value of the reading. Find the value of the constant n.

5) Two quantities t and x are plotted on log-log graph paper, t being plotted vertically and x being plotted horizontally. The result is a straight line and from the graph it is found that:

when $x = 8$, $t = 6.8$

and when $x = 20$, $t = 26.9$

Find the law connecting t and x.

6) The intensity of radiation, R, from certain radioactive materials at a particular time t is thought to follow the law $R = kt^n$.

In an experiment to test this the following values were obtained:

R	58	43.5	26.5	14.5	10
t	1.5	2	3	5	7

Show that the assumption was correct and evaluate the constants k and n.

7) The values given in the following table are thought to obey a law of the type $y = ab^{-x}$. Check this statement and find the values of the constants a and b.

x	0.1	0.2	0.4	0.6	1.0	1.5	2.0
y	175	158	60	32	6.4	1.28	0.213

8) The force F on the tight side of a driving belt was measured for different values of the angle of lap θ and the following results were obtained:

F	7.4	11.0	17.5	24.0	36.0
θ rad	$\pi/4$	$\pi/2$	$3\pi/4$	π	$5\pi/4$

Construct a graph to show these values conform approximately to an equation of the form $F = ke^{\mu\theta}$. Hence find the constants μ and k.

9) A capacitor and resistor are connected in series. The current i amperes after time t seconds is thought to be given by the equation $i = I.e^{-t/T}$ where I amperes is the initial charging current and T seconds is the time constant. Using the following values verify the relationship and find the values of the constants I and T:

i amperes	0.0156	0.0121	0.119 45	0.007 36	0.005 73
t seconds	0.05	0.10	0.15	0.20	0.25

SUMMARY

To verify the relationship $\quad z = a.t^n$

rewrite in the form $\quad \log z = n(\log t) + \log a$

and then plot $\log z$ against $\log t$ in order to obtain a straight line graph.

To verify the relationship $\quad z = a.b^t$

rewrite in the form $\quad \log z = (\log b)t + \log a$

and then plot $\log z$ against t in order to obtain a straight line graph.

To verify the relationship $\quad z = a.e^{bt}$

rewrite in the form $\quad \log z = (0.4343b)t + \log a$

and then plot $\log z$ against t in order to obtain a straight line graph.

 **PROBABILITY**

On reaching the end of this chapter you should be able to :-

1. Define probability and know that the total probability is unity.
2. Calculate simple probabilities.
3. Distinguish between dependent and independent events.
4. State the addition law of probability.
5. State the multiplication law of probability.
6. Calculate probabilities using the addition and multiplication laws.
7. Relate the binomial distribution to the expansion of $(q+p)^n$.
8. Calculate probabilities using the binomial theorem and tables of binomial coefficients.
9. Relate distributions of binomial probability to histogram representation.
10. Solve problems involving cumulation of terms of the binomial distribution.
11. Describe the conditions under which the binomial distribution approximates to the Poisson distribution.
12. Calculate probabilities for a Poisson distribution using tables of $e^{-\lambda}$.
13. Solve problems involving the cumulation of terms of the Poisson distribution.
14. Relate distributions of Poisson probability to histogram representation for given values of the variate.

SIMPLE PROBABILITY

If a coin is tossed it will come down either heads or tails. There are only these two possibilities. The probability of obtaining a head in a single toss is one possibility out of two possibilities. We write this:

$$\text{Pr (head)} = \frac{\text{possibility of a head}}{\text{total possibilities}} = \frac{1}{2}$$

When we roll a die (plural: dice) we can get one of six possible scores, 1, 2, 3, 4, 5 or 6. The probability of scoring 3 in a single roll of the die is one possibility out of a total of six possibilities. Hence:

$$\text{Pr (three)} = \frac{\text{possibility of a three}}{\text{total possibilities}} = \frac{1}{6}$$

There are 52 playing cards in a pack. When we cut the pack we can obtain one of these cards and hence the total possibilities are 52. Since there are four aces in the pack:

$$\text{Pr (ace)} = \frac{\text{possibility of an ace}}{\text{total possibilities}} = \frac{4}{52} = \frac{1}{13}$$

EXAMPLE 1

Calculate the probability of cutting a king, queen or jack in a single cut of a pack of cards.

Total possibilities = 52

Possibility of cutting a king, queen or jack = 12

$$\text{Pr (king, queen or jack)} = \frac{12}{52} = \frac{3}{13}$$

THE PROBABILITY SCALE

When an event is absolutely certain to happen we say that the probability of it happening is 1. Thus the probability that one day each of us will die is 1. When an event can never happen we say that the probability of it happening is 0. Thus the probability that any one of us can jump 5 metres high unaided is 0. All probabilities must, therefore, have a value between 0 and 1. They can be expressed as either a fraction or a decimal. Thus:

$$\text{Pr (head)} = \frac{1}{2} = 0.5$$

$$\text{Pr (ace)} = \frac{1}{13} = 0.077$$

Probabilities can be expressed on a probability scale (Fig. 15.1).

Any probability can be calculated from the formula:

$$p = \frac{\text{number of ways in which an event can happen}}{\text{total possibilities}}$$

EXAMPLE 2

In two successive throws of a die what is the probability that the sum of the scores will be eight?

If we examine the possible results we see that the following results can occur:

```
1, 1   2, 1   3, 1   4, 1   5, 1   6, 1
1, 2   2, 2   3, 2   4, 2   5, 2   6, 2
1, 3   2, 3   3, 3   4, 3   5, 3   6, 3
1, 4   2, 4   3, 4   4, 4   5, 4   6, 4
1, 5   2, 5   3, 5   4, 5   5, 5   6, 5
1, 6   2, 6   3, 6   4, 6   5, 6   6, 6
```

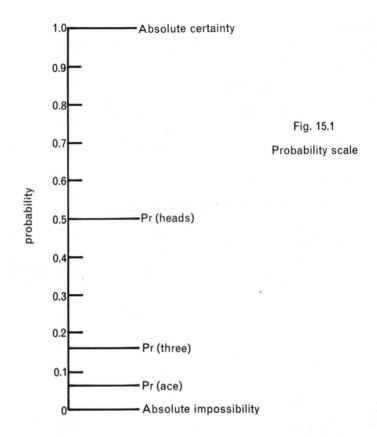

Fig. 15.1

Probability scale

We see that the total possibilities equal 36. The following results total eight: 2, 6; 3, 5; 4, 4; 5, 3; 6, 2. Hence the number of ways in which the event of scoring eight can happen is 5. Hence:

$$\text{Pr (eight)} = \frac{5}{36}$$

EMPIRICAL PROBABILITY

When rolling a die we expect each face to turn up in one-sixth of the number of throws. In any particular series of throws we would be rather surprised if we obtained exactly this result. However by making a large number of throws we shall get very near to one-sixth. This suggests another way of calculating the probability of an event happening.

$$p = \frac{\text{total number of the occurrences of the event}}{\text{total number of trials}}$$

The probability calculated by this formula is an empirical probability because it arises as a result of a statistical experiment.

For very many events it is not possible to obtain a theoretical probability and in such cases we resort to an empirical probability.

EXAMPLE 3

2000 parts were checked by measuring them and it was found that 100 were outside of the drawing limits (i.e., they were defective). What is the probability of finding a defective part in a single trial?

$$\text{Pr (defective part)} = \frac{\text{total number of defective parts found}}{\text{total number checked}}$$

$$= \frac{100}{2000} = 0.05$$

Probabilities are often expressed as percentages and in this case we would say:

$$\text{percentage defective} = 0.05 \times 100 = 5\%$$

TOTAL PROBABILITY

If p = the probability of an event happening

and q = the probability of it not happening

then $p+q = 1$.

That is,

the total probability, covering all possible events is always equal to 1.

Thus in tossing a coin if we regard a head as a success then $p = \frac{1}{2}$.
The event of obtaining a tail is then regarded as a failure and we say $q = \frac{1}{2}$. Thus the total probability, i.e., the probability of obtaining a head or a tail is $\frac{1}{2} + \frac{1}{2} = 1$.

Again, if in rolling a die we regard the rolling of a three as a success then $p = \frac{1}{6}$. Any other face of the die turning up will be regarded as a failure

and hence $q = \frac{5}{6}$.

The total probability covering all possible events is $\frac{1}{6} + \frac{5}{6} = 1$.

EXAMPLE 4

A bag contains 4 red balls, 3 blue, 2 green and one yellow. Find the probability of drawing a red ball in a single draw from the bag. What is the probability of not drawing a red ball?

$$\text{Pr (red)} = \frac{\text{number of ways of drawing a red ball}}{\text{total number of possibilities}} = \frac{4}{10} = 0.4$$

Since the probability of drawing a red ball is 0.4,

$$\text{Pr (not red)} = 1 - 0.4 = 0.6$$

Exercise 29

Find the probability of each of the following events occurring:

1) An ace, king, queen or jack of diamonds appearing when a pack of 52 playing cards is cut.

2) The sum of nine appearing when two dice are thrown simultaneously.

3) A silver coin being chosen at random from a purse containing 8 copper coins and 5 silver coins.

4) A bag contains 8 red, 5 white, 4 black and 3 green balls. A ball is drawn at random from the bag and replaced. Calculate the probability that it will be (a) red, (b) white, (c) not red, (d) not green, (e) black.

5) What is the probability that an even number will appear in a single roll of a die?

6) 1000 parts were checked by means of a limit gauge and 20 were found to fail the test. What is (a) the percentage defective, (b) the probability of finding a defective part in a single trial?

7) 20 ordinary bolts became accidently mixed with 180 high strength bolts. Calculate the probability of choosing a high strength bolt when one bolt is chosen at random from the 200 bolts.

ADDITION LAW OF PROBABILITY

If two or more events are such that not more than one of them can occur in a single trial they are said to be *mutually exclusive*.

Thus the events of throwing a 5 or a 6 in a single roll of a die are mutually exclusive events because it is only possible to throw either a 5 or a 6; it is impossible to throw both.

If p_1, p_2, p_3, ... are the separate probabilities of the occurrence of 1, 2, 3, ... mutually exclusive events then the probability that *one* of these events will occur is:

$$P = p_1 + p_2 + p_3 + \ldots$$

EXAMPLE 5

Find the probability of drawing either an ace, or the king of diamonds, or the queen of hearts, in a single cut of a pack of 52 playing cards.
In a single trial,

$$\text{Pr (ace)} = p_1 = \frac{4}{52} = \frac{1}{13}$$

$$\text{Pr (king of diamonds)} = p_2 = \frac{1}{52}$$

$$\text{Pr (queen of hearts)} = p_3 = \frac{1}{52}$$

$$\text{Pr (ace, king of diamonds or queen of hearts)} = p_1 + p_2 + p_3$$

$$= \frac{1}{13} + \frac{1}{52} + \frac{1}{52}$$

$$= \frac{6}{52} = 0.12$$

Note that the events of drawing an ace, the king of diamonds and the queen of hearts in a single cut are mutually exclusive events because only one of of three events can occur.

The addition law is sometimes called the *or law* because it is the probability of the occurrence of one event *or* another event that is required. In Example 5 it was the probability of drawing an ace *or* the king of diamonds *or* the queen of hearts that was required. Notice that when the addition law is used the chances of one of the events happening is *increased*.

EXAMPLE 6

It is known that the probability of obtaining 0 defectives in a sample of 40 items is 0.34, whilst the probability of obtaining 1 defective part is 0.46. What is the probability of obtaining:

(a) not more than 1 defective part in a sample,

(b) more than 1 defective part in a sample?

If we choose a random sample of 40 items then this sample may contain any number of defective parts in it up to a maximum of 40. The events of drawing samples with certain numbers of defectives in them are mutually exclusive events.

(a)   Pr (not more than 1 defective) = the probability of drawing a sample with 0 defective parts in it + the probability of drawing a sample with 1 defective part in it

$$= 0.34 + 0.46 = 0.80$$

(b)   Since the total probability covering all possible events is 1,

Pr (more than 1 defective) $= 1 - 0.80 = 0.20$

MULTIPLICATION LAW

Two or more events are said to be *independent* if the probability of the occurrence of any one of the events is not influenced by the occurrence of any other of the events.

In two separate cuts of a pack of cards, what happens on the first cut in no way affects what happens on the second cut. Hence these are independent events.

Similarly in two horse races, what happens in the first race in no way affects what happens in the second race and hence the two events are independent.

If $p_1, p_2, p_3, \ldots$ are the separate probabilities of the occurrence of 1, 2, 3, ... independent events the probability that *all* of the events will happen is:

$$P = p_1 \times p_2 \times p_3. \ldots$$

EXAMPLE 7

If three dice are thrown simultaneously, find the probability that each will show six.

The events of throwing the three dice are independent events and

on the first die,   Pr (six) $= p_1 = \dfrac{1}{6}$

on the second die,   Pr (six) $= p_2 = \dfrac{1}{6}$

on the third die,   Pr (six) $= p_3 = \dfrac{1}{6}$

$$\text{Pr (three sixes)} = p_1 \times p_2 \times p_3 = \frac{1}{6} \times \frac{1}{6} \times \frac{1}{6} = \frac{1}{216} = 0.0046$$

The multiplication law is sometimes called the *and law* because it is the probability of one event *and* another event which is required. In Example 7 it was the probability of throwing a six on the first die *and* throwing a six on the second die *and* throwing a six on the third die that was required. Notice that when the multipication law is used the chances that all the events will happen are reduced.

EXAMPLE 8

It is known that 10% of the items produced on a certain machine tool are defective. If a sample of three items is chosen at random from a large batch of these items, what is the probability that all three items will be defective?

Since what happens on the first draw in no way affects what happens on the other draws, the three events are independent.

On first draw,   Pr (defective) $= p_1 = 0.1$

On second draw,   Pr (defective) $= p_2 = 0.1$

On third draw,   Pr (defective) $= p_3 = 0.1$

$$\text{Pr (three defectives)} = p_1 \times p_2 \times p_3 = 0.1 \times 0.1 \times 0.1 = 0.001$$

It is possible to have a mixture of independent and dependent events. For instance, suppose in Example 8 we desire to know the probability of drawing only one defective part in a sample of three items.

There are three distinct possibilities as shown in the table below:

	1st draw	2nd draw	3rd draw
1st possibility	D	G	G
2nd possibility	G	D	G
3rd possibility	G	G	D

Any of these three possibilities gives only one defective in the sample. Each of the three possibilities are mutually exclusive events and hence:

$$\text{Pr (one defective)} = (0.1 \times 0.9 \times 0.9) + (0.9 \times 0.1 \times 0.9) + (0.9 \times 0.9 \times 0.1)$$

$$= 3 \times 0.1 \times (0.9)^2 = 0.243$$

Exercise 30

1) Find the probability of drawing either an ace, king or queen in a single draw from a pack of 52 playing cards.

2) The probability of obtaining 0 defective items in a sample is 0.20 whilst the probability of obtaining 1 defective part in the same size of sample is 0.25. Calculate the probability of drawing a sample with not more than one defective part in it.

3) A ball is drawn from a bag containing 3 red, 5 blue and 2 white balls. What is the probability that it will be either red or white?

4) Calculate the probability of rolling a 3 or a 6 in a single roll of a die.

5) If two fair dice are thrown at the same time, what is the probability of obtaining a score of 12?

6) Three ten pence coins are tossed together. Copy and complete the table below to show the ways in which the coins could land.

Coin 1	Coin 2	Coin 3
H	H	H

(a)   From your table write down the probability of:
    (i)   the three coins all falling tails,
    (ii) two coins falling heads and one coin tails.

(b)   The third coin is now changed for a six sided die and the two remaining coins and the die are tossed together. Calculate the probability that the throw will give:
    (i)   tail, tail, six,
    (ii) one head, one tail and either a 1 or a 2.

7) A penny, a halfpenny and a 10 p piece are tossed 200 times. For any particular toss find the probabilities that, assuming the coins are fair:

(a)   both copper coins are heads,

(b)   all three coins are heads,

(c)   at least one coin is heads,

(d)   at least two coins are heads.

8) A false coin is loaded so that on any one toss the probability of a head is $\frac{2}{3}$. The coin is tossed three times. Calculate the probabilities of obtaining:

(a)   three heads,

(b)   three tails,

(c)   2 tails,

(d)   2 heads.

9)   *A* and *B* are points in a toy train system (Fig. 15.2). The probability of going straight on at each point is $\frac{2}{3}$. Find the probability that:

(a)   the train *T* hits the waiting train,

(b)   the train *T* goes into the shed.

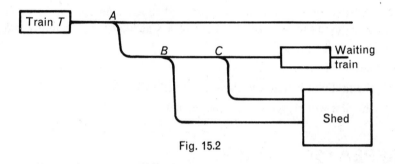

Fig. 15.2

10)   The pointer shown in Fig. 15.3 is spun. Find the probability that the pointer:

(a)   stops in section *B*,

(b)   stops in either *R* or *G* section.

The result of two successive spins are noted. Find the probability that:

(c)   the first spin is an *R* and the second spin is a *G*,

(d)   the pointer stops in either *R* then *G*   or   *G* then *R*.

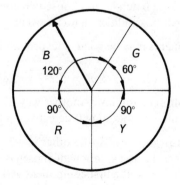

Fig. 15.3

11) A loaded die shows scores with the following probabilities:

Score	1	2	3	4	5	6
Probability	0.14	0.18	0.16	0.15	0.17	0.20

(a) On a single throw what is the probability of a score less than 3?

(b) If the die is thrown twice what is the probability of a 4 followed by a 6?

(c) The die is thrown twice and the two scores added together. What is the probability that the total score will be:

   (i) 2   (ii) 12   (iii) exactly 7   (iv) less than 5?

12) A bag contains 14 yellow counters and 10 red counters. Two counters are taken out in succession and not replaced. Calculate the probability that:

(a) both will be red,

(b) both will be yellow,

(c) the first will be red and the second yellow,

(d) the first will be yellow and the second red.

REPEATED TRIALS

Suppose that it is known that 10% of components produced on a certain machine are defective. If we choose one component at random from a large batch of these components then the probability of it being defective is $p = 10/100 = 0.1$.

The probability of it being good is $q = 90/100 = 0.9$.

Since the component will have to be classified as either good or defective we have $p + q = 1$.

Now from the same large batch let us choose two components at random. There are three distinct possibilities that can occur:

(a) Both the components may be good. The probability of this occurring is,

$$(q) \times (q) = q^2 = (0.9)^2 = 0.81$$

by using the multiplication law. Another way of saying this is to state that the probability of obtaining 0 defectives is 0.81.

(b) One component is good whilst the other is defective. This can occur in two ways: either the first component chosen is good with the second component defective, or the first component will be defective with the second good. By the multiplication law the probability of the first way occurring will be $q \times p = qp$, whilst the probability of the second way occurring will be $p \times q = pq$. If we are concerned only with the final result irrespective of the order in which the defectives occur then the probability of obtaining 1 defective is:

$$qp + pq = 2pq = 2 \times 0.1 \times 0.9 = 0.18$$

by using the addition law.

(c) Both components are defective. The probability of this occurring is:

$$(p) \times (p) = p^2 = (0.1)^2 = 0.01$$

Tabulating these results:

Possibility	Ways of arising	Probability of way	Probability of possibility occurring
0 defectives	good—good	q^2	q^2
1 defective	good—bad bad—good	qp pq	$2qp$
2 defectives	bad—bad	p^2	p^2

The terms in the table will be recognised as those arising from the expansion of $(q+p)^2$. Now let us analyse the results which will be obtained when taking a sample of 3 components.

The terms in the last column of the table will be recognised as those arising from the expansion of $(q+p)^3$.

Possibility	Ways of arising	Probability of way	Probability of possibility occurring
0 defectives	good good good	q^3	q^3
1 defective	good good defective good defective good defective good good	qqp qpq pqq	$3q^2p$
2 defectives	good defective defective defective good defective defective defective good	qpp pqp ppq	$3qp^2$
3 defectives	defective defective defective	ppp	p^3

From the foregoing we will rightly expect that the probabilities of obtaining 0, 1, 2, 3 and 4 defectives in a sample of 4 components would be the successive terms of the expansion of $(q+p)^4$.

Now,

$$(q+p)^4 = q^4 + 4q^3p + 6q^2p^2 + 4qp^3 + p^4$$

Thus, if $p = 0.1$ and $q = 0.9$:

the probability of 0 defectives $= q^4 = (0.9)^4 = 0.6561$

the probability of 1 defective $= 4q^3p = 4 \times (0.9)^3 \times (0.1) = 0.2916$

the probability of 2 defectives $= 6q^2p^2 = 6 \times (0.9)^2 \times (0.1)^2 = 0.0486$

the probability of 3 defectives $= 4qp^3 = 4 \times (0.9) \times (0.1)^3 = 0.0036$

the probability of 4 defectives $= p^4 = (0.1)^4 = 0.0001$

the total probability covering all the possible results $= 1.0000$

We can now deduce the general rule.

The probabilities of obtaining 0, 1, 2, 3, . . . n defectives in a sample of n items is given by the successive terms of the expansion of $(q+p)^n$.

In using this general rule it must be clearly understood that:

n = number in the sample,

p = probability of finding a defective in a single trial,

q = probability of finding a good item in a single trial,

and

$$p+q = 1$$

The expansion of $(q+p)^n$ can be obtained either by using the binomial theorem (see Chapter 12) or the table of binomial coefficients given below.

BINOMIAL COEFFICIENTS

The following table gives the binomial coefficients. The values are as found using Pascal's triangle but arranged differently.

The extreme left hand column is the value of n in $(q+p)^n$, and the top row is the number, r, of the particular term in the expression.

$r:$	0	1	2	3	4	5	6	7	8	9	10
$n =$											
1	1	1									
2	1	2	1								
3	1	3	3	1							
4	1	4	6	4	1						
5	1	5	10	10	5	1					
6	1	6	15	20	15	6	1				
7	1	7	21	35	35	21	7	1			
8	1	8	28	56	70	56	28	8	1		
9	1	9	36	84	126	126	84	36	9	1	
10	1	10	45	120	210	252	210	120	45	10	1
11	1	11	55	165	330	462	462	330	165	55	11
12	1	12	66	220	495	792	924	792	495	220	66
13	1	13	78	286	715	1287	1716	1716	1287	715	286
14	1	14	91	364	1001	2002	3003	3432	3003	2002	1001
15	1	15	105	455	1365	3003	5005	6435	6435	5005	3003
16	1	16	120	560	1820	4368	8008	11440	12870	11440	8008
17	1	17	136	680	2380	6188	12376	19448	24310	24310	19448
18	1	18	153	816	3060	8568	18564	31824	43758	48620	43758
19	1	19	171	969	3876	11628	27132	50388	75582	92378	92378
20	1	20	190	1140	4845	15504	38760	77520	125970	167960	184756

Thus for example $(a+b)^5 = a^5+5a^4b+10a^3b^2+10a^2b^3+5ab^4+b^5$.

Note that the coefficients are symmetrical. Thus:

$$(a+b)^{15} = a^{15}+15a^{14}b+105a^{13}b^2+455a^{12}b^3+1365a^{11}b^4+3003a^{10}b^5$$

$$+5005a^9b^6+6435a^8b^7+6435a^7b^8+5005a^6b^9+3003a^5b^{10}$$

$$+1365a^4b^{11}+455a^3b^{12}+105a^2b^{13}+15ab^{14}+b^{15}.$$

EXAMPLE 9

It is known that 10% of certain articles manufactured are defective. Find the probabilities of obtaining 0, 1, 2, 3, 4 and 5 defectives in a sample of 5 articles taken at random from a large batch of these articles.

The probability of a single article, chosen at random, being defective is $p = 0.1$. The probability of it being satisfactory is $q = 1-0.1 = 0.9$. Now:

$$(q+p)^5 = q^5+5q^4p+10q^3p^2+10q^2p^3+5qp^4+p^5$$

Number of Defectives	Term of Expansion	Probability	
0	q^5	$(0.9)^5$	$= 0.590\ 49$
1	$5q^4p$	$5 \times (0.9)^4 \times (0.1)$	$= 0.328\ 05$
2	$10q^3p^2$	$10 \times (0.9)^3 \times (0.1)^2$	$= 0.072\ 90$
3	$10q^2p^3$	$10 \times (0.9)^2 \times (0.1)^3$	$= 0.008\ 10$
4	$5qp^4$	$5 \times (0.9) \times (0.1)^4$	$= 0.000\ 45$
5	p^5	$(0.1)^5$	$= 0.000\ 01$

Total probability covering all possible results $= 1.000\ 00$

These results are shown in histogram form in Fig. 15.4.

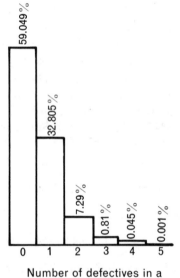

Number of defectives in a sample of five articles

Fig. 15.4.

EXAMPLE 10

A process is known to produce 5% of faulty articles. Estimate the chance of a sample of 12 articles (a) containing exactly 3 faulty articles, (b) containing more than 3 faulty articles.

Here $p = 0.05$ and $q = 0.95$. We need the first 4 terms of the expansion of $(q+p)^{12}$.

Number of defectives	Term of expansion	Probability
0	q^{12}	$(0.95)^{12}$ $= 0.5400$
1	$12q^{11}p$	$12 \times (0.95)^{11} \times (0.05)$ $= 0.3411$
2	$66q^{10}p^2$	$66 \times (0.95)^{10} \times (0.05)^2 = 0.0987$
3	$220q^9p^3$	$220 \times (0.95)^9 \times (0.05)^3 = 0.0173$
		Total probabilities of above cases $= 0.9971$

(a) The probability of obtaining exactly 3 defectives is 0.0173 or 1.73%.

(b) Since the total probability is 1, the probability of obtaining more than 3 defectives is $1 - 0.9971 = 0.0029$ or 0.29%.

THE BINOMIAL DISTRIBUTION

When a frequency distribution is obtained by use of the expansion of $(q+p)^n$ it is called a binomial distribution. By using the successive terms of $(q+p)^n$ a theoretical frequency table can be drawn up. A histogram can then be drawn which will show the distribution.

EXAMPLE 11

Draw a histogram for the expected frequencies obtained in tossing 10 coins 1024 times.

Here $p = q = \frac{1}{2}$, and $1024 = 2^{10}$

$$(q+p)^{10} = q^{10} + 10q^9p + 45q^8p^2 + 120q^7p^3 + 210q^6p^4 + 252q^5p^5$$

$$+ 210q^4p^6 + 120q^3p^7 + 45q^2p^8 + 10qp^9 + p^{10}$$

The histogram for the frequency table shown on p. 199 is shown in Fig. 15.5 and it is seen to be symmetrical. A histogram for a binomial distribution is symmetrical if, and only if, $p = q = \frac{1}{2}$. In Fig. 15.4 the histogram is skewed to one side, the values of p and q being 0.1 and 0.9 respectively.

Number of Heads	Term from expansion	Frequency
0	$(\frac{1}{2})^{10} = \dfrac{1}{1024}$	1
1	$10 \times (\frac{1}{2})^9(\frac{1}{2}) = \dfrac{10}{1024}$	10
2	$45 \times (\frac{1}{2})^8(\frac{1}{2})^2 = \dfrac{45}{1024}$	45
3	$120 \times (\frac{1}{2})^7(\frac{1}{2})^3 = \dfrac{120}{1024}$	120
4	$210 \times (\frac{1}{2})^6(\frac{1}{2})^4 = \dfrac{210}{1024}$	210
5	$252 \times (\frac{1}{2})^5(\frac{1}{2})^5 = \dfrac{252}{1024}$	252
6	$210 \times (\frac{1}{2})^4(\frac{1}{2})^6 = \dfrac{210}{1024}$	210
7	$120 \times (\frac{1}{2})^3(\frac{1}{2})^7 = \dfrac{120}{1024}$	120
8	$45 \times (\frac{1}{2})^2(\frac{1}{2})^8 = \dfrac{45}{1024}$	45
9	$10 \times (\frac{1}{2})(\frac{1}{2})^9 = \dfrac{10}{1024}$	10
10	$(\frac{1}{2})^{10} = \dfrac{1}{1024}$	1

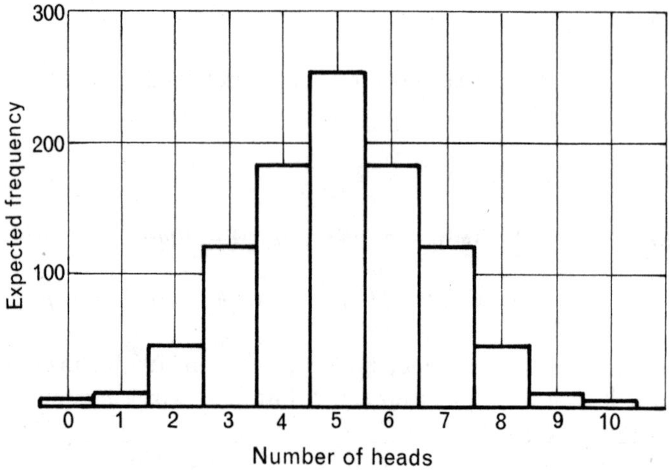

Fig. 15.5

It can be shown that if np is less than 5 the histogram showing the expected frequency distribution will be skewed. If np is greater than 5 the histogram is reasonably symmetrical.

Exercise 31

1) It is known that in a large quantity of articles 10% are defective. Find the probabilities of obtaining 0, 1, 2, 3, or 4 defectives in a sample of 4 articles.

2) A machine is known to produce 5% of faulty articles. Find the probability of a sample of 20 articles containing: (a) no faulty articles, (b) more than one faulty article.

3) A class of 20 students is prepared for an examination. Experience shows that 30% of the students fail. Find the probability of exactly 6 failing the examination.

4) A product is being made in large quantities. Successive samples of 40 items give the following numbers of defectives: 2, 2, 0, 0, 1, 3, 0, 2, 0, 0. Find the probability of obtaining in a sample of 40 items: (a) no defectives, (b) one defective, (c) more than one defective.

5) It is known that 10% of resistors produced are defective. What is the probability that in a sample of 12 such resistors more than 2 will be defective?

6) A section in a certain factory consists of 30 workpeople. Over a long period of time the average number of absentees per shift is 2. Find the probability of more than 2 absentees in any single shift.

7) In a mass production process the average number of defectives produced is 5%. 500 samples each consisting of 20 articles are examined. Show that the expected number of defectives is as shown:

Number of defectives	0	1	2	3	4	5
Frequency	180	189	94	31	4	5

8) In a large batch of fuses it is known that 10% are faulty. Show that in a batch of 500 samples each containing 8 fuses the expected frequencies are:

Number of defectives	0	1	2	3	4	5
Frequency	215	191	75	17	2	0

9) Samples of 6 items are drawn from a production process in which it is known that 10% of the articles made are defective. Draw a histogram showing the probabilities of 0, 1, 2, 3, 4, 5, and 6 defectives in a sample.

THE POISSON DISTRIBUTION

In all cases of repeated trials the binomial theorem may be used to calculate the probabilities. However, as the value of n (the sample size) increases the amount of work involved in calculating the probabilities becomes increasingly great.

In most sampling schemes used in industry, a high degree of accuracy in calculating the probabilities is not essential and approximations to the binomial distribution are therefore used.

One of the approximations used is the Poisson distribution. This is used when:

(a) The probability of the event occurring in a single trial is small, i.e., not greater than 0.1.

(b) The number in the sample is greater than 30.

(c) The expectation, i.e., np, remains constant and is less than 5.

These conditions are almost always satisfied in practical industrial situations. Now there exists a series for e^λ which is:

$$e^\lambda = 1+\lambda+\frac{\lambda^2}{2!}+\frac{\lambda^3}{3!}+\ldots$$

where e is the base of natural logarithms. Its value, correct to 4 places of decimals, is 2.7183.

In any calculation involving probabilities it is imperative that the total probability, covering all possible events, is equal to 1.

Now $\qquad\qquad e^\lambda \times e^{-\lambda} = e^0 = 1$

Hence we can use the product $e^\lambda \times e^{-\lambda}$ to form a theoretical frequency distribution when it is written:

$$e^{-\lambda}\left(1+\lambda+\frac{\lambda^2}{2!}+\frac{\lambda^3}{3!}+\ldots\right)$$

A distribution obtained by using this series is called a *Poisson distribution* and it forms a close approximation to the Binomial distribution provided the conditions given above apply and provided $\lambda = np$.

The successive terms of the series give the probabilities of there being 0, 1, 2, ... defectives in the sample. Thus:

Number of defective items in the sample	Probability of finding the stated number of defective items in the sample
0	$e^{-\lambda}$
1	$(e^{-\lambda})\lambda$
2	$e^{-\lambda}\left(\dfrac{\lambda^2}{2!}\right)$
3	$e^{-\lambda}\left(\dfrac{\lambda^3}{3!}\right)$ etc.

Values of $e^{-\lambda}$ are given in most books of mathematical tables, or may be obtained using a scientific calculator.

EXAMPLE 12

A machine is known to produce 5% of faulty articles. A sample of 4 items is drawn at random from a large batch of these articles. Find the probabilities of obtaining 0, 1, 2, 3 and 4 defectives in the sample using: (a) the binomial distribution, (b) the Poisson distribution.

(a) For the binomial distribution, $p = 0.05$, $q = 0.95$ and $n = 4$. The expansion of $(q+p)^4$ is,

$$q^4 + 4q^3p + 6q^2p^2 + 4qp^3 + p^4$$

Number of defectives	Term of the expansion	Probability
0	q^4	$(0.95)^4 = 0.814\ 51$
1	$4q^3p$	$4 \times (0.95)^3 \times (0.05) = 0.171\ 48$
2	$6q^2p^2$	$6 \times (0.95)^2 \times (0.05)^2 = 0.013\ 54$
3	$4qp^3$	$4 \times (0.95) \times (0.05)^3 = 0.000\ 46$
4	p^4	$(0.05)^4 = 0.000\ 01$

Total probability covering all possible cases $= 1.000\ 00$

(b) For the Poisson distribution, $p = 0.05$, $n = 4$ and $\lambda = np = 0.2$.
$$e^{-\lambda} = e^{-0.2} = 0.8187$$

Number of defectives	Term in the series	Probability	
0	$e^{-\lambda}$		0.8187
1	$(e^{-\lambda})\lambda$	0.8187×0.2	$= 0.1637$
2	$e^{-\lambda}\left(\dfrac{\lambda^2}{2!}\right)$	$0.8187 \times \dfrac{(0.2)^2}{2 \times 1}$	$= 0.0164$
3	$e^{-\lambda}\left(\dfrac{\lambda^3}{3!}\right)$	$0.8187 \times \dfrac{(0.2)^3}{3 \times 2 \times 1}$	$= 0.0012$
4	$e^{-\lambda}\left(\dfrac{\lambda^4}{4!}\right)$	$0.8187 \times \dfrac{(0.2)^4}{4 \times 3 \times 2 \times 1}$	$= 0.0000$
		Total probability covering all possible cases	$= 1.0000$

Comparing the results given by the binomial and Poisson distributions we see that there are small discrepancies in the probabilities as given by the two distributions. For most applications these are too small to be significant and we can use the Poisson distribution as an approximation to the binomial distribution *provided* we adhere to the conditions given previously.

EXAMPLE 13

A process is known to produce 5% of faulty articles. A sample of 40 such articles is taken. Find: (a) the chance of obtaining exactly 4 defectives in the sample, (b) the chance of obtaining more than 4 defectives in the sample. (c) Represent the probabilities of there being 0, 1, 2, 3 and 4 faulty articles in a histogram.

Here $p = 0.05$ and $\lambda = np = 0.05 \times 40 = 2$
$$e^{-\lambda} = e^{-2} = 0.1353$$

Number of defectives	Term in the series	Probability	
0	$e^{-\lambda}$		0.1353
1	$e^{-\lambda}(\lambda)$	0.1353×2	$= 0.2706$
2	$e^{-\lambda}\left(\dfrac{\lambda^2}{2!}\right)$	$0.1353 \times \dfrac{2^2}{2 \times 1}$	$= 0.2706$
3	$e^{-\lambda}\left(\dfrac{\lambda^3}{3!}\right)$	$0.1353 \times \dfrac{2^3}{3 \times 2 \times 1}$	$= 0.1804$
4	$e^{-\lambda}\left(\dfrac{\lambda^4}{4!}\right)$	$0.1353 \times \dfrac{2^4}{4 \times 3 \times 2 \times 1}$	$= 0.0902$
		Total	$= 0.9471$

(a)   The chance of obtaining 4 defectives is 0.0902.

(b)   Since the total probability is unity and the sum of the probabilities of obtaining 0, 1, 2, 3 and 4 defectives is 0.9471, the probability of obtaining more than four defectives is,

$$1-0.9471 = 0.0529$$

(c)   The histogram is shown in Fig. 15.6.

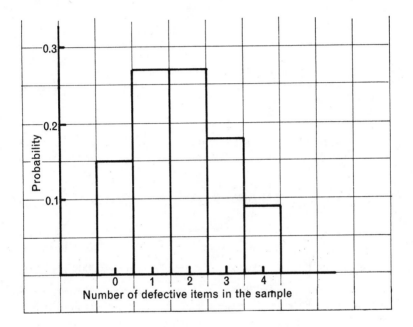

Fig. 15.6

EXAMPLE 14

In the mass production of an article 1000 samples each of 50 articles are examined. The numbers of defectives are shown in the following frequency table. Find the average number of defectives per sample and hence show that the distribution is approximately the same as the Poisson distribution with this average.

Defectives	0	1	2	3	4	5
Frequency	674	261	57	7	1	0

Total number of defectives $= (261 \times 1) + (57 \times 2) + (7 \times 3) + (1 \times 4)$

$$= 261 + 114 + 21 + 4$$

$$= 400$$

Total number of articles $= 50 \times (674 + 261 + 57 + 7 + 1)$

$$= 50 \times 1000 = 50\,000$$

$\therefore$ $$p = \frac{400}{50\,000} = 0.008$$

$\therefore$ $$\lambda = np = 50 \times 0.008 = 0.4$$

$\therefore$ $$e^{-\lambda} = e^{-0.4} = 0.6703$$

Number of defectives	Term in the series	Probability		Frequency
0	$e^{-\lambda}$		$= 0.6703$	670.3
1	$(e^{-\lambda})\lambda$	0.6703×0.4	$= 0.2681$	268.1
2	$e^{-\lambda}\left(\dfrac{\lambda^2}{2!}\right)$	$0.6703 \times \dfrac{(0.4)^2}{2 \times 1}$	$= 0.0536$	53.6
3	$e^{-\lambda}\left(\dfrac{\lambda^3}{3!}\right)$	$0.6703 \times \dfrac{(0.4)^3}{3 \times 2 \times 1}$	$= 0.0071$	7.1
4	$e^{-\lambda}\left(\dfrac{\lambda^4}{4!}\right)$	$0.6703 \times \dfrac{(0.4)^4}{4 \times 3 \times 2 \times 1}$	$= 0.0007$	0.7
5	$e^{-\lambda}\left(\dfrac{\lambda^5}{5!}\right)$	$0.6703 \times \dfrac{(0.4)^5}{5 \times 4 \times 3 \times 2 \times 1}$	$= 0.0000$	0

The comparison between the values calculated by the Poisson distribution agree very well with those given in the frequency table.

THE POISSON DISTRIBUTION IN ITS OWN RIGHT

We have seen that we may use the Poisson distribution as an approximation to the binomial distribution. We can do this provided we know the value of n (the sample size) and p (the fraction defective) in the expression $(q+p)^n$. There are, however, very many cases where n is not known. For instance, in checking the number of surface flaws on sheets of plastic or in checking the number of defective items in batches of unknown size.

In such cases we may use the Poisson distribution to calculate the probabilities provided that λ is made equal to the average value of the occurrence of the event.

EXAMPLE 15

By checking several boxes containing large numbers of fuses it was found that the average number of defective fuses per box was 3. Find the probability of finding a box containing 4 or more defective fuses.

Here we have $\lambda = 3$ and $e^{-3} = 0.0498$.

Number of defective fuses in the sample	Probability	
0	$e^{-\lambda}$	$= 0.0498$
1	$(-e^{\lambda})\lambda = 0.0498 \times 3$	$= 0.1494$
2	$e^{-\lambda}\left(\dfrac{\lambda^2}{2!}\right) = 0.0498 \times \dfrac{3^2}{2 \times 1}$	$= 0.2241$
3	$(e^{-\lambda})\left(\dfrac{\lambda^3}{3!}\right) = 0.0498 \times \dfrac{3^3}{3 \times 2 \times 1}$	$= 0.2241$
		0.6474

Hence the probability of finding a box containing 4 or more defective fuses is $p = 1 - 0.6474 = 0.3526$. We mean that 35.26% of the boxes are likely to contain 4 or more defective fuses.

EXAMPLE 16

20 sheets of aluminium alloy were examined for surface flaws. The number of flaws per sheet were as follows:

Sheet number	1 2 3 4 5 6 7 8 9 10 11 12 13 14 15 16 17 18 19 20
Number of flaws	2 0 1 3 2 1 0 0 1 0 2 2 1 0 2 2 1 0 3 2

Find the probability of finding a sheet, chosen at random from a large batch, which has 2 or more surface flaws.

Now $$\lambda = \frac{\text{total number of flaws in the 20 sheets}}{20}$$

or $$\lambda = \frac{25}{20} = 1.25$$

$\therefore$ $$e^{-\lambda} = 0.2865$$

Number of flaws per sheet	Probability
0	$e^{-\lambda} = 0.2865$
1	$(e^{-\lambda})\lambda = 0.3581$
	0.6446

Hence the probability of finding a sheet having 2 or more surface flaws is $p = 1 - 0.6446 = 0.3554$. This probability indicates that it is likely that 35.5% of the sheets will have 2 or more surface flaws.

Exercise 32

1)   It is known that 10% of the items produced on a certain machine are defective. Using the Poisson distribution calculate the probabilities of obtaining 0, 1, 2, 3 or 4 defective items in a sample of 4 items.

2)   1% of articles produced in a mass production process are known to be defective. If these are packed in boxes containing 300 articles find the probability that the box will contain exactly 3 defective articles.

3)   In a large consignment of resistors the average number in a box that were faulty was 2. Calculate the probability of: (a) finding a box containing exactly 2 defective resistors, (b) finding a box containing 2 or more defective resistors.

4)   From the production line of a certain factory 10 successive samples each containing 100 items were checked. The number of items failing the test were: 1, 2, 2, 4, 2, 1, 0, 1, 2, 0. Estimate the probability that a sample of 100 items taken at random will contain 2 or more defective items.

5)   10 sheets of plastic were examined for flaws. The number of flaws per sheet were as follows:

Sheet number	1	2	3	4	5	6	7	8	9	10
No. of flaws	5	3	2	6	4	3	5	2	5	5

Calculate the percentage of sheets which are likely to have 4 or more flaws.

6)   Over a long period of time, records show that on the average 3 people in a small factory are absent from a certain shift. Find the probability that: (a) everyone will be present on the shift, (b) that less than 2 will be absent, (c) 3 or more will be absent.

7)   A class of 15 students is prepared for an examination. Experience shows that 10% will obtain a distinction. What is the probability that 1 or more of this class will obtain a distinction in the examination.

8)   A pipe line in a chemical processing factory contains a great many valves. It is known that, on average, 2 of these valves will leak. What is the probability that at any given time less than 2 valves will leak?

SUMMARY

a) Simple probability:

$$p = \frac{\text{the number of ways in which an event can happen}}{\text{the total number of ways that are possible}}$$

b) The total probability $= 1$.

c) Addition law of probability. If two or more events are such that not more than one of them can occur in a single trial the events are said to be mutually exclusive. If $p_1, p_2, p_3, \ldots$ are the separate probabilities of the occurrence of 1, 2, 3, ... mutually exclusive events then the probability of one of the events occurring is $p = p_1 + p_2 + p_3 + \ldots$

d) Multiplication law of probability. Two or more events are independent if the probability of any one of the events occurring is not influenced by the occurrence of any other of the events. If $p_1, p_2, p_3, \ldots$ are the separate probabilities of the occurrence of 1, 2, 3, . . . independent events then the probability of all these events occurring is
$$p = p_1 \times p_2 \times p_3 \times \ldots.$$

e) If p is the fraction defective and q is the fraction good then in a sample of n items the probabilities of the sample containing 0, 1, 2, 3, ... defective items is given by the successive terms of the expansion of $(q+p)^n$. Note that $p+q = 1$.

f) If $\lambda = np$, where n is the sample size and p is the fraction defective then the probabilities of the sample containing 0, 1, 2, 3, ... defective items is given by the successive terms of the Poisson distribution:

$$e^{-\lambda}\left(1+\lambda+\frac{\lambda^2}{2!}+\frac{\lambda^3}{3!}+\ldots\right)$$

g) If n is not known then $\lambda = $ the average occurrence of the event, and the Poisson distribution may then be used in the same ways as in **f**.

 # THE NORMAL DISTRIBUTION

On reaching the end of this chapter you should be able to :-

1. Calculate the probabilities for normal distributions with mean zero and standard deviation of unity, using normal distribution tables.

2. Convert data from a general normal distribution to standardised form.

3. Solve problems concerning the normal distribution using 2 and tables.

4. Plot relative frequency percentages against variate on normal probability paper.

5. Determine whether points lie on a straight line in 4.

6. Calculate mean and standard deviation using normal probability tables.

REPRESENTATION OF FREQUENCY DISTRIBUTIONS

There are two ways whereby similar articles produced by a process may be checked. They are:

(a) By using limit gauges which merely classify the articles as good or defective.

(b) By measuring the characteristics of the article (e.g., lengths, diameters, etc.) to see if they conform to the drawing specification.

It will be recalled that a frequency distribution may be represented by a frequency curve. If limit gauges are used then the frequency distribution will approximate to a Poisson distribution and the frequency curve will be skewed as shown in Fig. 16.1.

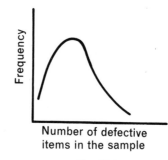

Fig. 16.1

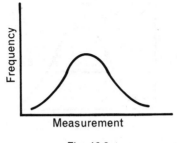

Fig. 16.2

When the data is obtained by actual measurement the frequency curve approximates to the symmetrical bell shaped curve shown in Fig. 16.2. This will only be so if sufficient measurements are made and usually 100 will suffice.

This bell shaped curve is called the *normal distribution curve* and it can be defined in terms of the total frequency, the arithmetic mean and the standard deviation.

ARITHMETIC MEAN AND STANDARD DEVIATION

Since the normal curve is symmetrical about its centre-line, the centre-line represents the mean of the distribution. Hence the mean locates the position of the curve from the reference axis as shown in Fig. 16.3 which displays similar distributions with different means.

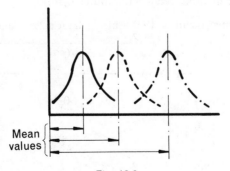

Fig. 16.3

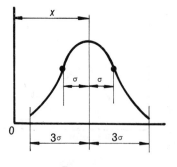

Fig. 16.4

The standard deviation, σ, gives a measure of the spread (or dispersion) of the curve about the mean. As shown in Fig. 16.4, the standard deviation is the distance from the mean to the points of inflexion (points where the curve bends back on itself) of the curve. Although the normal curve extends to infinity on either side of the mean, for most practical purposes it may be assumed to terminate at three standard deviations on either side of the mean.

The method of calculating the mean and standard deviation was dealt with in Level II of the course. However the example which follows may be used to revise the method.

EXAMPLE 1

Calculate the mean and standard deviation for the distribution shown below:

Diameter (mm)	18.3–18.5	18.6–18.8	18.9–19.1	19.2–19.4	19.5–19.7
Frequency	6	25	37	25	7

Using the coded method, the calculation is as follows:

Assumed mean = 19.0 mm   Unit size = 0.3 mm

Class	x	x_c	f	fx_c	fx_c^2
18.3–18.5	18.4	−2	6	−12	24
18.6–18.8	18.7	−1	25	−25	25
18.9–19.1	19.0	0	37	0	0
19.2–19.4	19.3	1	25	25	25
19.5–19.7	19.6	2	7	14	28

$\Sigma f = 100$ $\Sigma fx_c = 2$ $\Sigma fx_c^2 = 102$

$$\bar{x}_c = \frac{\Sigma f x_c}{\Sigma f} = \frac{2}{100} = 0.02$$

$\bar{x}$ = assumed mean $+ \bar{x}_c \times$ unit size

$= 19.0 + 0.02 \times 0.3 = 19.006$ mm

$$\sigma_c = \sqrt{\frac{\Sigma f x_c^2}{\Sigma f} - (\bar{x}_c)^2} = \sqrt{\frac{102}{100} - 0.02^2} = \sqrt{1.02 - 0.0004}$$

$$= \sqrt{1.0196} = 1.0097$$

$\sigma = \sigma_c \times$ unit size $\quad = 1.0097 \times 0.3 = 0.3029$ mm

PROBABILITY FROM THE NORMAL CURVE

Since the total probability must always be equal to 1, if we make the area under the normal curve equal to 1, then we can use areas under the curve to estimate probabilities.

For convenience, the origin is taken at the intersection of the mean and the horizontal axis (Fig. 16.5). The horizontal axis is then marked off in units of the standard deviation. Any deviation from the mean can then be calculated by using the formula:

$$u = \frac{x - \bar{x}}{\sigma}$$

as shown in Fig. 16.6.

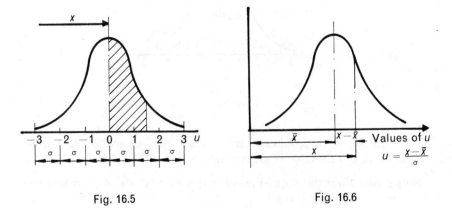

Fig. 16.5   Fig. 16.6

EXAMPLE 2

For a certain process it is found that $\bar{x} = 15.00$ mm and $\sigma = 0.30$ mm. Express the measurements: (a) 14.70 mm and (b) 15.21 mm in terms of the standard deviation.

(a) Here $x = 14.70$ mm.

$$\therefore \qquad u = \frac{14.70 - 15.00}{0.30} = -1$$

Hence the measurement of 14.70 mm lies 1 standard deviation below the mean.

Note that when x is less than $\bar{x}$ the value of u is negative.

(b) Here $x = 15.21$ mm.

$$u = \frac{15.21 - 15.00}{0.30} = 0.7$$

Hence the measurement 15.21 mm lies 0.7 of a standard deviation above the mean.
Note that when x is greater than $\bar{x}$ the value of u is positive.

The shaded area (Fig. 16.5) gives the probability of something occurring between the mean value $(u = 0)$ and 1.5 standard deviations from the mean $(u = 1.5)$.

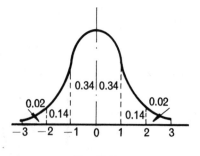

Fig. 16.7

Fig. 16.7 shows approximate areas under the normal curve but for accurate values the table of areas under the normal curve should be used.

Notice that since the normal curve is symmetrical about its centre-line, the area of each half is 0.5.

AREAS UNDER THE STANDARD NORMAL CURVE FROM 0 TO u

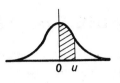

u	0	1	2	3	4	5	6	7	8	9
0.0	.0000	.0040	.0080	.0120	.0160	.0199	.0239	.0279	.0319	.0359
0.1	.0398	.0438	.0478	.0517	.0557	.0596	.0636	.0675	.0714	.0754
0.2	.0793	.0832	.0871	.0910	.0948	.0987	.1026	.1064	.1103	.1141
0.3	.1179	.1217	.1255	.1293	.1331	.1368	.1406	.1443	.1480	.1517
0.4	.1554	.1591	.1628	.1664	.1700	.1736	.1772	.1808	.1844	.1879
0.5	.1915	.1950	.1985	.2019	.2054	.2088	.2123	.2157	.2190	.2224
0.6	.2258	.2291	.2324	.2357	.2389	.2422	.2454	.2486	.2518	.2549
0.7	.2580	.2612	.2642	.2673	.2704	.2734	.2764	.2794	.2823	.2852
0.8	.2881	.2910	.2939	.2967	.2996	.3023	.3051	.3078	.3106	.3133
0.9	.3159	.3186	.3212	.3238	.3264	.3289	.3315	.3340	.3365	.3389
1.0	.3413	.3438	.3461	.3485	.3508	.3531	.3554	.3577	.3599	.3621
1.1	.3643	.3665	.3686	.3708	.3729	.3749	.3770	.3790	.3810	.3830
1.2	.3849	.3869	.3888	.3907	.3925	.3944	.3962	.3980	.3997	.4015
1.3	.4032	.4049	.4066	.4082	.4099	.4115	.4131	.4147	.4162	.4177
1.4	.4192	.4207	.4222	.4236	.4251	.4265	.4279	.4292	.4306	.4319
1.5	.4332	.4345	.4357	.4370	.4382	.4394	.4406	.4418	.4429	.4441
1.6	.4452	.4463	.4474	.4484	.4495	.4505	.4515	.4525	.4535	.4545
1.7	.4554	.4564	.4573	.4582	.4591	.4599	.4608	.4616	.4625	.4633
1.8	.4641	.4649	.4656	.4664	.4671	.4678	.4686	.4693	.4699	.4706
1.9	.4713	.4719	.4726	.4732	.4738	.4744	.4750	.4756	.4761	.4767
2.0	.4772	.4778	.4783	.4788	.4793	.4798	.4803	.4808	.4812	.4817
2.1	.4821	.4826	.4830	.4834	.4838	.4842	.4846	.4850	.4854	.4857
2.2	.4861	.4864	.4868	.4871	.4875	.4878	.4881	.4884	.4887	.4890
2.3	.4893	.4896	.4898	.4901	.4904	.4906	.4909	.4911	.4913	.4916
2.4	.4918	.4920	.4922	.4925	.4927	.4929	.4931	.4932	.4934	.4936
2.5	.4938	.4940	.4941	.4943	.4945	.4946	.4948	.4949	.4951	.4952
2.6	.4953	.4955	.4956	.4957	.4959	.4960	.4961	.4962	.4963	.4964
2.7	.4965	.4966	.4967	.4968	.4969	.4970	.4971	.4972	.4973	.4974
2.8	.4974	.4975	.4976	.4977	.4977	.4978	.4979	.4979	.4980	.9481
2.9	.4981	.4982	.4982	.4983	.4984	.4984	.4985	.4985	.4986	.4986
3.0	.4987	.4987	.4987	.4988	.4988	.4989	.4989	.4989	.4990	.4990
3.1	.4990	.4991	.4991	.4991	.4992	.4992	.4992	.4992	.4993	.4993
3.2	.4993	.4993	.4994	.4994	.4994	.4994	.4994	.4995	.4995	.4995
3.3	.4995	.4995	.4995	.4996	.4996	.4996	.4996	.4996	.4996	.4997
3.4	.4997	.4997	.4997	.4997	.4997	.4997	.4997	.4997	.4997	.4998
3.5	.4998	.4998	.4998	.4998	.4998	.4998	.4998	.4998	.4998	.4998
3.6	.4998	.4998	.4999	.4999	.4999	.4999	.4999	.4999	.4999	.4999
3.7	.4999	.4999	.4999	.4999	.4999	.4999	.4999	.4999	.4999	.4999
3.8	.4999	.4999	.4999	.4999	.4999	.4999	.4999	.4999	.4999	.4999
3.9	.5000	.5000	.5000	.5000	.5000	.5000	.5000	.5000	.5000	.5000

USING THE TABLE OF AREAS

Fig. 16.8 shows a typical area. To use the table first note that the figures in the first column are values of u in increments of 0.1. The corresponding value in the column headed 0 gives the area between $u = 0$ and the value of u in column one.

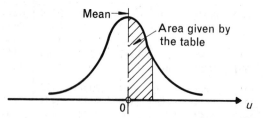

Fig. 16.8

Thus the area between $u = 0$ and $u = 0.7$ is 0.2580. If the value of u has two decimal places the area will be found in the appropriate column of the next nine columns. Thus the area between $u = 0$ and $u = 0.74$ is 0.2704.

If the value of u is negative, the table is used in exactly the same way and in reading the table the negative sign is ignored.

The following examples will show how the table is used:

EXAMPLE 3

Find the area under the normal curve between:

(a) $u = 0$ and $u = 0.63$ (b) $u = 0$ and $u = -0.85$

(c) $u = 1.2$ and $u = 1.73$ (d) $u = -0.63$ and $u = 1.12$

(a)   The area between $u = 0$ and $u = 0.63$ is as shown in Fig. 16.9 and is found directly from the table as 0.2357

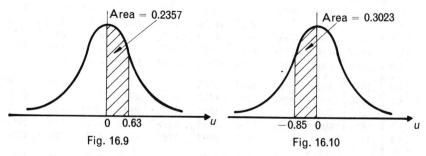

Fig. 16.9   Fig. 16.10

(b)   Since the normal curve is symmetrical about the mean we can find the required area between $u = 0$ and $u = -0.85$ directly from the table as if it was that between $u = 0$ and $u = 0.85$. Its value is 0.3023 (see Fig. 16.10).

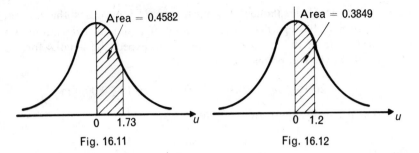

Fig. 16.11   Fig. 16.12

(c) First find the area between $u = 0$ and $u = 1.73$. This is 0.4582 and is shown in Fig. 16.11. Next find the area between $u = 0$ and $u = 1.2$. This is 0.3849 and is shown in Fig. 16.12.

By subtracting the area 0.3849 from the area 0.4582 we obtain the required area, 0.0733, which is shown in Fig. 16.13.

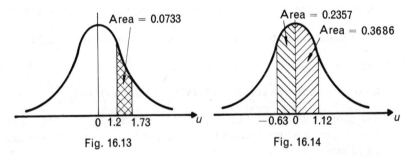

Fig. 16.13   Fig. 16.14

(d) First we find the area between $u = 0$ and $u = -0.63$, which is the same as the area between $u = 0$ and $u = 0.63$. Its value is 0.2357. Next find the area between $u = 0$ and $u = 1.12$, which is 0.3686.

The required area is the sum of these two, as shown in Fig. 16.14, giving a value of 0.6043.

EXAMPLE 4

By measuring a large number of components produced on an automatic lathe it was found that the mean length of the components was 20.10 mm with a standard deviation of 0.03 mm. Find:

(a) Within what limits you would expect the lenghts for the whole of the components to lie.

(b) The probability that one component taken at random would have: (i) a length between 20.05 mm and 20.12 mm, (ii) a length less than 20.02 mm, (iii) a length greater than 20.17 mm.

(a) As previously explained, for most practical purposes the normal curve may be regarded as terminating at 3 standard deviations on either side of the mean. Thus, we would expect the lengths for the whole of the component to lie between:

$$20.10 \pm 3\sigma = 20.10 \pm 3 \times 0.03 = 20.10 \pm 0.09 \text{ mm}$$

that is between 20.01 mm and 20.19 mm.

(b) (i)
$$u_1 = \frac{20.05 - 20.10}{0.03} = -1.67$$

$$u_2 = \frac{20.12 - 20.10}{0.03} = 0.67$$

using the table of areas under the normal curve:

Area under the curve between $u = 0$ and $u = -1.67 = 0.4525$.

Area under the curve between $u = 0$ and $u = 0.67 = 0.2486$.

Hence between $u = 0$ and $u = 0.67$

total area under the curve $= 0.4525 + 0.2486 = 0.7011$.

The probability is 0.7011 or 70.11%. This indicates that 70.11% of all the components are expected to have a length between 20.05 mm and 20.12 mm.

(ii)
$$u = \frac{20.02 - 20.10}{0.03} = -2.67$$

Now between $u = 0$ and $u = -2.67$

the area under curve $= 0.4962$.

Since the normal curve is symmetrical and since its area is unity, the total area to the left of the mean is 0.5000. The probability of obtaining a component with a length less than 20.02 mm is therefore $0.5000 - 0.4962 = 0.0038$ or 0.38%. This means that 0.38% of all components produced are expected to have a length less than 20.02 mm.

(iii)
$$u = \frac{20.17 - 20.1}{0.03} = 2.33$$

Now between $u = 0$ and $u = 2.33$

the area under curve $= 0.4901$.

The probability of obtaining a component with a length greater than 20.17 mm therefore, $0.5000 - 0.4901 = 0.0099$ or 0.99%. This means that 0.99% of all the components produced are expected to have a length greater than 20.17 mm.

EXAMPLE 5

10 000 resistors are to be produced with a specification limit of 100 ± 7 ohms. By measuring the first 100 made it was found that their mean resistance was 102 ohms with a standard deviation of 5 ohms. Determine how many of these resistors are likely to be outside the specification limits.

The first step is to calculate the value of u at the lower specification limit.

$$u_1 = \frac{93 - 102}{5} = -1.8$$

The area between $u = 0$ and $u = -1.8$ is 0.4641. Hence the tail area to the left of $u = -1.8$ is $0.5 - 0.4641 = 0.0359$ or 3.59% (Fig. 16.15). Hence the number of resistors likely to be produced outside of the lower specification limit is 3.59% of 10 000 = 359.

We next calculate the value of u at the upper sepcification limit.

$$u_2 = \frac{107 - 102}{5} = 1$$

The area between $u = 0$ and $u = 1$ is 0.3413. The tail area to the right of $u = 1$ is $0.5 - 0.3413 = 0.1587$ or 15.87%. Hence the number of resistors likely to be produced outside of the upper specification limit is 15.87% of 10 000 = 1587.

The total number of resistors likely to be produced outside of specification limits is $359 + 1587 = 1946$.

Note that in Fig. 16.15 it is the shaded tail areas which give the probabilities of a resistor being produced which is outside of the specification limits.

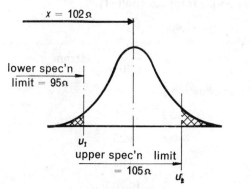

Fig. 16.15

EXAMPLE 6

On checking a large number of steel bars it was found that 5% were under 132.7 mm in length. 35% were between 132.7 mm and 135.5 mm. Find the mean and standard deviation assuming a normal distribution.

The given percentages are represented as areas in Fig. 16.16. We find that for an area of 0.45, $u = 1.65$, and for an area of 0.10, $u = 0.25$, and substituting in the equation: $u = \dfrac{x - \bar{x}}{\sigma}$

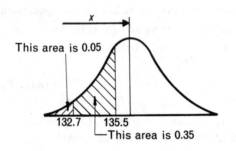

This area is 0.05

132.7   135.5

└─This area is 0.35

Fig. 16.16

Then $$-1.65 = \frac{132.7 - \bar{x}}{\sigma}$$

or $$-1.65\sigma = 132.7 - \bar{x} \qquad [1]$$

And $$-0.25 = \frac{135.5 - \bar{x}}{\sigma}$$

or $$-0.25\sigma = 135.5 - \bar{x} \qquad [2]$$

Subtracting equation [2] from equation [1],

$$-1.40\sigma = -2.8$$

$\therefore$ $$\sigma = \frac{2.8}{1.40} = 2.00 \text{ mm}$$

Substituting for σ in equation [2],

$$-0.25 \times 2 = 135.5 - \bar{x}$$

$\therefore$ $$-0.5 = 135.5 - \bar{x}$$

$\therefore$ $$\bar{x} = 135.5 + 0.5$$

$\therefore$ $$\bar{x} = 136.0 \text{ mm}$$

THE NORMAL DISTRIBUTION AS AN APPROXIMATION TO THE BINOMIAL DISTRIBUTION FOR REPEATED TRIALS

It can be shown that, for a binomial distribution:

the mean $\qquad\qquad\qquad\qquad \bar{x} = np$

and the standard deviation $\qquad\quad \sigma = \sqrt{npq}$

where $\quad p =$ probability of success in a single trial,

$\qquad\quad q =$ probability of failure in a single trial $= 1-p$

and $\quad\;\; n =$ number of trials.

This approximation should only be used when np is greater than 5.

EXAMPLE 7

A manufacturer of light bulbs finds that on the average 3% are defective. What is the probability that 40 or more will be defective in 1000 bulbs selected at random?

$$p = 0.03 \quad q = 0.97$$

$$\bar{x} = np = 1000 \times 0.03 = 30$$

$$\therefore \qquad \sigma = \sqrt{npq} = \sqrt{1000 \times 0.03 \times 0.97}$$

$$= \sqrt{29.1} = 5.394$$

$$\therefore \qquad u = \frac{x - \bar{x}}{\alpha} = \frac{40 - 30}{5.394} = \frac{10}{5.394} = 1.85$$

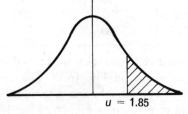

$$u = 1.85$$

Fig. 16.17

We now need to find the shaded area of Fig. 16.17. The area between $u = 0$ and $u = 1.85$ is 0.4678. The probability is, therefore,

$$0.5000 - 0.4678 = 0.0322 \quad \text{or} \quad 3.22\%$$

CHECKING A DISTRIBUTION TO SEE IF IT IS A NORMAL DISTRIBUTION

One way of checking a distribution to see if it is a normal distribution is to use probability paper. The method is shown in Example 8.

EXAMPLE 8

Check if the distribution given below is a normal distribution.

Variate	2.26	2.27	2.28	2.29	2.30	2.31	2.32	2.33	2.34
Frequency	1	3	25	60	126	61	20	3	1

The first step is to convert the given frequency distribution into a cumulative frequency distribution. Remembering that the upper boundary limit of the first class is 2.265, the cumulative frequency distribution is then as follows:

Variate	Cumulative frequency	Percentage cumulative frequency
less than 2.265	1	$\frac{1}{300} \times 100 = 0.33\%$
less than 2.275	4	$\frac{4}{300} \times 100 = 1.33\%$
less than 2.285	29	$\frac{29}{300} \times 100 = 8.17\%$
less than 2.295	89	29.7%
less than 2.305	215	71.7%
less than 2.315	276	90.2%
less than 2.325	296	98.7%
less than 2.335	299	99.7%
less than 2.345	300	100%

The percentage cumulative frequencies are plotted against the variate on the probability paper shown in Fig. 16.18. The closeness to which the plotted points conform to a straight line determines whether the distribution is nearly normal or not. From Fig. 16.18 we shall see that the given distribution may, for all practical purposes, be accepted as a normal distribution.

The mean and standard deviation may also be found from Fig. 16.18. For a normal distribution the mean is the value of the variate corresponding to a cumulative percentage frequency of 50%. From the diagram the mean of the distribution is found to be about 2.30.

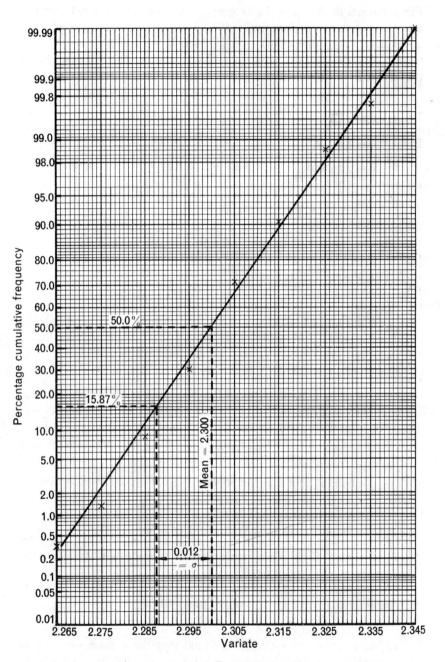

Fig. 16.18

For a normal distribution the area under the normal curve between the mean and 1 standard deviation is 0.3413 or 34.13% (see table of areas under the normal curve). Now   50%−34.13% = 15.87%.   Value of the variate corresponding to   15.87% = 2.288.   Hence the standard deviation is   2.30−2.288 = 0.012.

Exercise 33

In the following, $u = (x-\bar{x})/\sigma$.

1)   If $\bar{x} = 30$ and $\sigma = 5$ find the values of u corresponding to:
(a) $x = 22$ (b) $x = 28$ (c) $x = 35$ (d) $x = 41$

2)   Find the areas under the normal curve:
(a)   between $u = 0$ and $u = 1.20$
(b)   between $u = -0.75$ and $u = 0$
(c)   between $u = -0.3$ and $u = 1.12$
(d)   between $u = -2.3$ and $u = -1.56$
(e)   between $u = 0.58$ and $u = 2.41$
(f)   less than $u = -0.79$
(g)   greater than $u = 2.11$
(h)   greater than $u = -0.86$
(i)   less than $u = 1.12$
(j)   less than $u = 2.49$

3)   By measuring a large number of components produced on an automatic lathe it was found that the mean length of the components was 18.60 mm with a standard deviation of 0.02 mm. Find:

(a)   within what limits the lengths of the whole of the components could be expected to lie.

(b)   the probability that one component taken at random would have:
    (i)   a length less than 18.56 mm,
    (ii)   a length greater than 18.65 mm,
    (iii)   a length between 18.59 mm and 18.62 mm.

4)   For a normal distribution with a mean of 20 and a standard deviation of 4 find the probabilities for:
(a)   the interval 12 to 18,   (d)   28 and above,
(b)   the interval 17 to 25,   (e)   14 and less.
(c)   the interval 15 to 30,

5)   In mass producing a bush it is found that the average diameter is 12.5 mm with a standard deviation of 0.015 mm. If 2000 bushes are to be produced find the number of bushes that can be expected to have dimensions between 12.475 and 12.525 mm.

6) An article is being produced on an automatic lathe. 300 of these are measured and a certain dimension is checked to the nearest 0.01 mm. The results are given in the following table:

Dimension	9.96	9.97	9.98	9.99	10.00	10.01	10.02	10.03	10.04
Frequency	1	6	25	72	93	69	27	6	1

Find the probability that one article chosen at random would have:

(a) a dimension less than 9.975 mm,
(b) a dimension greater than 10.035 mm,
(c) a dimension between 9.985 and 10.015 mm.

7) In checking the heights of a large group of men it is found that 8% are under 1.55 m in height and 40% have a height between 1.55 m and 1.68 m. Asssuming that a normal distribution holds find the mean and standard deviation of the group. What percentage of the group may be expected to have a height greater than 1.73 m.

8) The masses of a group of 200 castings are given in the table below:

Mass (kg)	168	168.5	169	169.5	170	170.5	171	171.5	172
Frequency	2	2	28	44	50	38	26	6	4

Find the probability of the mass of a single casting falling between 168.75 kg and 171.25 kg.

9) An article is being mass produced. The drawing demands that a certain dimension shall be 12.51 ± 0.04 mm. 600 of the articles are measured with the following results:

Dimension (mm)	12.46	12.47	12.48	12.49	12.50	12.51	12.52	12.53	12.54
Frequency	1	7	52	121	250	123	40	5	1

Calculate the mean and standard deviation. Hence find how many of the 600 articles are likely to meet the drawing specification.

10) In a certain factory 1000 electric lamps are installed. These lamp have an average burning life of 1100 hours, and a standard deviation of 250 hours. How many lamps are likely to fail during the first 800 burning hours and how many lamps are likely to fail between 900 and 1400 burning hours?

11) 1000 resistors are to be made with specification limits of 33 ± 0.6 ohm. It is known that the mean of the resistors will be 33.2 ohm with a standard deviation of 0.3 ohm. Find how many of the 1000 resistors will be produced outside of the specification limits.

12) As a result of testing 300 oil filled capacitors the following frequency distribution was obtained.

Capacity (μF)	19.96	19.97	19.98	19.99	20.00	20.01	20.02	20.03	20.04
Frequency	1	6	25	72	93	69	27	6	1

By using probability paper establish if the distribution is approximately normal or not and estimate the mean and standard deviation.

13) The masses of 200 castings were as shown in the following table:

Mass (kg)	168	168.5	169	169.5	170	170.5	171	171.5	172
Frequency	2	2	28	44	50	38	26	6	4

By using probability paper show that the distribution is very nearly normal and hence determine the mean and standard deviation.

14) A manufacturer knows that 2% of his products are, on the average, defective. What is the probability that 1000 articles will contain 18 or more defectives.

15) The average life of a certain battery is 24 months with a standard deviation of 6 months. If 800 batteries are to be sent to a distributor how many can be expected to fail between 12 and 18 months?

16) In a mass production process it is known that 2% of the articles produced are defective. What is the probability of obtaining 22 or more defectives in a sample of 800 such articles?

17) When samples of 1500 articles are examined it is found that on the average 15 articles are defective. In a sample of 1500 articles what is the probability that it will contain between 13 and 18 defectives?

SUMMARY

a) The normal curve is obtained when the data is obtained by measurement as opposed to using limit gauges, etc.

b) When the area under the normal curve is made equal to unity, the curve can be used for estimating probabilities.

c) When the origin is taken as the intersection of the mean and the horizontal axis and the horizontal axis is marked off in units of the standard deviation any deviation from the mean can be calculated from the formula:

$$u = \frac{x - \bar{x}}{\sigma}$$

d) For accurate values of areas under the normal curve the table of areas under the normal curve should be used. Since the normal curve is symmetrical about the mean, the area under each half of the curve is 0.5.

e) The normal distribution may be used as an approximation to the binomial distribution provided the expectation, np, is greater than 5. In such cases:

$$\bar{x} = np \quad \text{and} \quad \sigma = \sqrt{npq}$$

where p = the probability of a success in a single trial

q = the probability of failure in a single trial = $1-p$

and n = the number of trials.

f) To check if a distribution is approximately normal, plot the percentage cumulative frequency against the variate on probability paper. If the points lie approximately on a straight line the distribution is approximately normal. The mean is then the value of the variate corresponding to a percentage cumulative frequency of 50%. The value of the standard deviation is the difference between the value of the variate corresponding to a cumulative frequency of 50% and the value of the variate corresponding to a cumulative frequency of 15.87%.

MOMENTS OF INERTIA

On reaching the end of this chapter you should be able to :-

1. Determine that the moment of inertia $I = \Sigma \, mr^2$ for a system of concentrated rotating masses.
2. Define radius of gyration k from the expression $I = (\Sigma \, m)k^2$.
3. Deduce units of moment of inertia.
4. Understand that all the mass in an elementary ring may be considered to be concentrated at one radius — hence its polar $I = Mr^2$.
5. Use the above information to find by integration the polar I of a solid cylinder.
6. State the parallel axis theorem.
7. Apply the above to typical engineering component shapes.

MOMENT OF INERTIA

Moment of inertia is the property of a body used in rotational problems just as the mass is used in problems involving linear motion.

The symbol for moment of inertia is I and must always be stated together with the reference axis about which it has been calculated, in a similar manner as was used for second moment of area. It is unfortunate that engineers have chosen the same symbol for moment of inertia as for second moment of area but, in practice, confusion hardly ever occurs.

The linear kinetic energy of a mass m moving with a velocity v is given by $\frac{1}{2}mv^2$.

Also, the tangential velocity v at the end of a radius r rotating about an axis through O, and perpendicular to the plane of the paper, with angular velocity ω (Fig. 17.1) is given by $v = r\omega$.

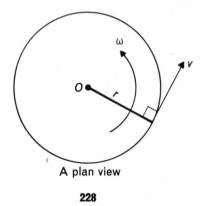

Fig. 17.1

A plan view

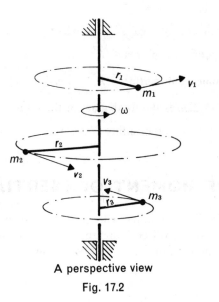

A perspective view

Fig. 17.2

Consider the system shown in Fig. 17.2 which comprises three concentrated masses m_1, m_2, and m_3 each being fixed to the end of a radius arm. The radius arms are all attached to a vertical spindle, the arms and spindle being assumed to have negligible mass. The whole system rotates with an angular velocity ω.

$$\text{The kinetic energy of the system} = \frac{1}{2}m_1v_1^2 + \frac{1}{2}m_2v_2^2 + \frac{1}{2}m_3v_3^2$$

$$= \frac{1}{2}m_1(r_1\omega)^2 + \frac{1}{2}m_2(r_2\omega)^2 + \frac{2}{1}m_3(r_3\omega)^2$$

$$= \frac{1}{2}(m_1r_1^2 + m_2r_2^2 + m_3r_3^2)\omega^2$$

$$= \frac{1}{2}\left(\sum mr^2\right)\omega^2$$

$$= \frac{1}{2}I\omega^2$$

where **the moment of inertia of the system is** $I = \Sigma\, mr^2$.

Although the expression $I = \Sigma\, mr^2$ has been derived for three rotating masses it is true for a system comprising any number of concentrated masses.

This expression for I may be used, together with integration, for finding the moment of inertia of bodies whose mass is *not* concentrated at any one particular radius. Example 3 shows this method.

Now for first moments of area $\qquad (\Sigma\ A)\bar{x} = (\Sigma\ Ax)$

and for second moments of area $\quad I = (\Sigma\ A)k^2 = (\Sigma\ Ax^2)$

Similarly **for moments of inertia:** $\quad \boxed{I = (\Sigma\ m)k^2 = (\Sigma\ mr^2)}$

where $\qquad\qquad \Sigma\ m$ is the total mass of the body

and $\qquad\qquad\qquad k$ is the radius of gyration

UNITS OF MOMENT OF INERTIA

Since $I = (\Sigma\ m)k^2$ the units will be those of mass $\times$ distance^2. In basic SI units the mass will be kilogrammes and the distance metres, and so the units of moment of inertia will be **kg m^2**.

EXAMPLE 1

Find the moment of inertia of a component about an axis, given that its mass if 5 kg and the radius of gyration about this axis is 200 mm.

Now $\qquad\qquad\qquad I = $ total mass $\times k^2$

$$= 5 \times \left(\frac{200}{1000}\right)^2$$

$$= 0.2 \text{ kg m}^2$$

EXAMPLE 2

Find the polar moment of inertia of a rim type flywheel whose dimensions are as shown in Fig. 17.3. The density of the steel from which it is made is 7800 kg/m^3.

A rim type flywheel is a wheel in which all of its mass may be considered to be concentrated round the rim, the hub and spokes being neglected in calculations.

Hence all the mass may be considered to be situated at one particular radius, namely the mean radius of the rim and this may be taken as the radius of gyration.

Hence $\qquad\qquad$ polar $k = 0.6 - 0.04 = 0.56$ m

The volume of the rim $= 2\pi \times$ (mean radius) $\times$ (width) $\times$ (thickness)

$$= 2\pi \times 0.56 \times 0.25 \times 0.08$$

$$= 0.0704 \text{ m}^3$$

and   mass of the rim = volume × density

$$= 0.0704 \times 7800$$

$$= 549 \text{ kg}$$

Hence   polar I = total mass × k^2

$$= 549 \times 0.56^2$$

$$= 172 \text{ kg m}^2$$

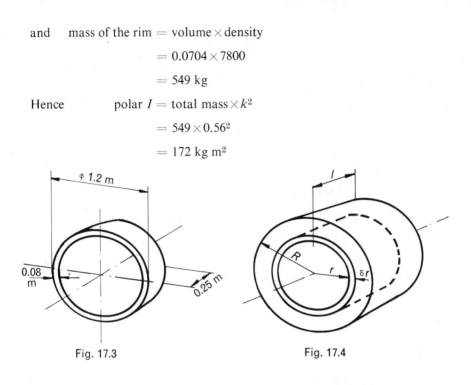

Fig. 17.3   Fig. 17.4

EXAMPLE 3

Find the moment of inertia, about its polar axis, of a solid cylinder of mass M and radius R.

The mass of a solid cylinder is *not* concentrated at a particular radius and so we will consider that the cylinder is made up from a series of elementary rings (each similar to the rim of a rim type flywheel), Fig. 17.4.

Let the density of the material be ρ, and the length of the cylinder be l.

Now for the cylinder $I = \Sigma\,(mr^2)$

$$= \Sigma\,(\text{mass of elementary ring} \times r^2)$$

$$= \Sigma\,(\text{volume of elementary ring} \times \text{density} \times r^2)$$

$$= \Sigma(2\pi r\ \delta r\ l\rho r^2)$$

$$= \Sigma\,(2\pi\rho l r^3\ \delta r)$$

$$= \int_0^R 2\pi\rho l r^3\ dr$$

$$= 2\pi\rho l \int_0^R r^3\ dr$$

$$= 2\pi \rho l \left[\frac{r^4}{4} \right]_0^R$$

$$= \frac{2\pi \rho l R^4}{4}$$

$$= \frac{1}{2}(\pi R^2 l \rho) R^2$$

$$= \frac{1}{2} M R^2$$

since the mass of the cylinder is $M = \pi R^2 l \rho$.

If we also wished to find the radius of gyration k about the polar axis we may use: $\qquad\qquad I = Mk^2$ since $M = \Sigma m$

$$\therefore \qquad\qquad \frac{1}{2}MR^2 = Mk^2$$

from which $\qquad\qquad k^2 = \frac{R^2}{2}$

or $\qquad\qquad k = \frac{R}{\sqrt{2}}$

PARALLEL AXIS THEOREM

This is similar to that for second moments of area.

In Fig. 17.5 the axis CD is parallel to the axis through centre of gravity G, h being the distance between the axes. Then:

$$\boxed{I_{CD} = I_{\text{axis thro' }G} + Mh^2}$$

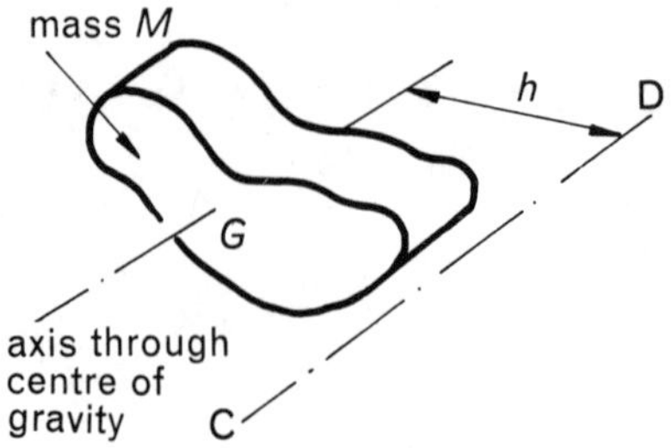

mass M

axis through
centre of
gravity   C

Fig. 17.5

EXAMPLE 4

Find the moment of inertia for the connecting rod shown in Fig. 17.6, about an axis through its centre of gravity G and parallel to the knife edge. Its mass is 41 kg and I about the knife edge is 15.5 kg m² found by oscillating the connecting-rod as a compound pendulum.

Using the parallel axis theorem:

$$I_{\text{knife edge}} = I_{\text{axis thro'}G} + Mh^2$$

$$\therefore \quad 15.5 = I_{\text{axis thro'}G} + 41\left(\frac{530}{1000}\right)^2$$

$$\therefore \quad 15.5 = I_{\text{axis thro'}G} + 11.5$$

$$\therefore \quad I_{\text{axis thro' }G} = 15.5 - 11.5$$

$$= 4 \text{ kg m}^2$$

Moments of inertia of various solids may be found in most engineering reference books, typical data being as shown below.

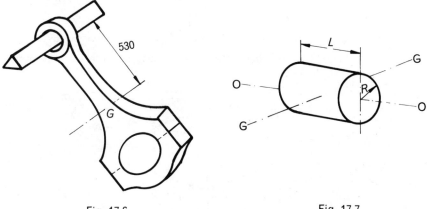

Fig. 17.6   Fig. 17.7

Solid cylinder (Fig. 17.7)

$$\text{Polar } I_{OO} = \frac{1}{2}MR^2$$

$$I_{GG} = M\left(\frac{L^2}{12} + \frac{R^2}{4}\right)$$

where GG passes through the centre of gravity and is perpendicular to OO.

Solid rectangular block   (Fig. 17.8)

$$I_{GG} = M\left(\frac{a^2+b^2}{12}\right)$$

where GG passes through the centre of gravity and is perpendicular to face *ab*.

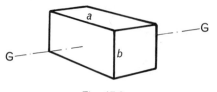

Fig. 17.8

Exercise 34

1)   Find the polar moment of inertia of a solid aluminium cylinder 100 mm diameter and 500 mm long, if the density of aluminium is 2700 kg/m³.

2)   Find the radius of gyration k_{GG} of the solid rectangular block shown in Fig. 17.9 if axis GG is perpendicular to the face measuring 200 mm × 300 mm and passes through the centre of gravity of the block.

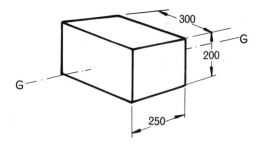

Fig. 17.9

3)   Find the polar moment of inertia of a hollow steel cylinder, 750 mm long, having an outside diameter of 500 mm and an inside diameter of 400 mm. The density of steel is 7900 kg/m³.

(*Hint:*   The polar moment of inertia of the hollow cylinder is the difference between the polar moments of inertia of a solid cylinder 500 mm diameter and a solid cylinder 400 mm diameter.)

4) A hollow copper cylinder is as shown in Fig. 17.10. If the density of copper is 9000 kg/m³ find:

(a) I_{OO}

(b) I_{AA}

(*Hint:* when using the parallel axis theorem which gives $I_{AA} = I_{OO} + Mh^2$ remember that M is the mass of the hollow cylinder.)

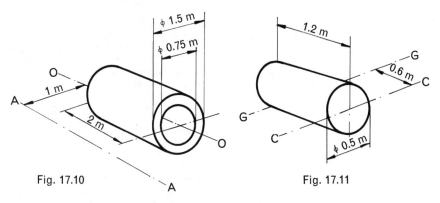

Fig. 17.10   Fig. 17.11

5) Fig. 17.11 shows a solid copper cylinder which has a density of 9000 kg/m³. Find:

(a) I_{GG}, where GG is an axis passing through the centre of gravity and perpendiculat to the polar axis.

(b) I_{CC}

6) Find I_{BB} for the solid block shown in Fig. 17.12 if the density of the material from which it is made is 8000 kg/m³.

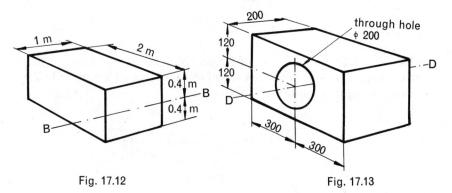

Fig. 17.12   Fig. 17.13

7) The component shown in Fig. 17.13 is made from a material which has a density of 7500 kg/m³. Find I_{DD}.

 # COMPLEX NUMBERS

On reaching the end of this chapter you should be able to :-

1. Understand the necessity of extending the number system to include the square roots of negative numbers.
2. Define j as $\sqrt{-1}$.
3. Define a complex number as consisting of a real part and an imaginary part.
4. Define a complex number z in the algebraic form $x+jy$.
5. Determine the complex roots of $ax^2+bx+c= 0$ when $b^2 < 4ac$ using the quadratic formula.
6. Perform the addition and subtraction of complex numbers in algebraic form.
7. Define the conjugate of a complex number in algebraic form.
8. Perform the multiplication and division of complex numbers in algebraic form.
9. Represent the algebraic form of a complex number on an Argand diagram, and show how it may be represented as a phasor.
10. Deduce that j may considered to be an operator, such that when the phasor representing $x+jy$ is multiplied by j it rotates the phasor through 90° anti-clockwise.
11. Understand how phasors on an Argand diagram may be added and subtracted in a manner similar to the addition and subtraction of vectors.
12. Show that the full polar form of a complex number is $(\cos\theta+j\sin\theta)$ which may be abbreviated to r/θ.
13. Perform the operations involved in the conversion of complex numbers in algebraic form to polar form and vice-versa.
14. Multiply and divide numbers in polar form.
15. Apply the above to problem arising from relevant engineering technology.

COMPLEX NUMBERS

The solution of the quadratic equation $ax^2+bx+c = 0$ is given by the

formula:
$$x = \frac{-b\pm\sqrt{b^2-4ac}}{2a}$$

When we use this formula most of the quadratic equations we meet, when solving engineering problems, are found to have roots which are ordinary positive or negative numbers.

Consider now the equation $x^2-4x+13 = 0$

then
$$x = \frac{-(-4)\pm(\sqrt{-4)^2-4\times1\times13}}{2\times1}$$

$$= \frac{4\pm\sqrt{16-52}}{2}$$

$$= \frac{4\pm\sqrt{-36}}{2}$$

236

$$= \frac{4 \pm \sqrt{(-1)(36)}}{2}$$

$$= \frac{4 \pm \sqrt{(-1)} \times \sqrt{(36)}}{2}$$

$$= \frac{4 \pm \sqrt{(-1)} \times 6}{2}$$

$$= 2 \pm \sqrt{-1} \times 3$$

It is not possible to find the value of the square root of a negative number.

In order to try to find a meaning for roots of this type we represent $\sqrt{-1}$ by the symbol j.

(Books on pure mathematics often use the symbol i, but in engineering j is preferred as i is used for the instantaneous value of a current.)

Thus the roots of the above equation become $2+j3$ and $2-j3$.

Definition of a complex number

Expressions such as $2+j3$ are called *complex numbers*. The number 2 is called the *real part* and $j3$ is called the *imaginary part*.

The general expression for a complex number is $x+jy$, which has a real part equal to x and an imaginary part equal to jy. The form $x+jy$ is said to be *the algebraic form* of a complex number. It may also be called *the cartesian form* or *rectangular notation*.

Powers of j

We have defined j such that

$$j = \sqrt{-1}$$

$\therefore$ squaring both sides of the equation

$$j^2 = (\sqrt{-1})^2 = -1$$

Hence $\qquad j^3 = j^2 \times j = (-1) \times j = -j$

and $\qquad j^4 = (j^2)^2 = (-1)^2 = 1$

and $$j^5 = j^4 \times j = 1 \times j = j$$

and $$j^6 = (j^2)^3 = (-1)^3 = -1$$

$$\text{and so on.}$$

The most used of the above relationships is $j^2 = -1$.

Addition and subtraction of complex numbers in algebraic form

The real and imaginary parts must be treated separately. The real parts may be added and subtracted and also the imaginary parts may be added and subtracted, both obeying the ordinary laws of algebra.

Thus
$$(3+j2)+(5+j6) = 3+j2+5+j6$$
$$= (3+5)+j(2+6)$$
$$= 8+j8$$

and
$$(1-j2)-(-4+j) = 1-j2+4-j$$
$$= (1+4)-j(2+1)$$
$$= 5-j3$$

EXAMPLE 1

If z_1, z_2 and z_3 represent three complex numbers such that $z_1 = 1.6+j2.3$, $z_2 = 4.3-j0.6$ and $z_3 = -1.1-j0.9$ find the complex numbers which represent:

(a) $z_1+z_2+z_3$

(b) $z_1-z_2-z_3$

(a)
$$z_1+z_2+z_3 = (1.6+j2.3)+(4.3-j0.6)+(-1.1-j0.9)$$
$$= 1.6+j2.3+4.3-j0.6-1.1-j0.9$$
$$= (1.6+4.3-1.1)+j(2.3-0.6-0.9)$$
$$= 4.8+j0.8$$

(b)
$$z_1-z_2-z_3 = (1.6+j2.3)-(4.3-j0.6)-(-1.1-j0.9)$$
$$= 1.6+j2.3-4.3+j0.6+1.1+j0.9$$
$$= (1.6-4.3+1.1)+j(2.3+0.6+0.9)$$
$$= -1.6+j3.8$$

Multipliciation of complex numbers in algebraic form

Consider the product of two complex numbers, $(3+j2)(4+j)$.

The brackets are treated in exactly the same way as the rules of algebra

make $(a+b)(c+d) = ac+bc+ad+bd$

Hence $(3+j2)(4+j) = 3\times4+j2\times4+3\times j+j2\times j$

$$= 12+j8+j3+j^22$$

$$= 12+j8+j3-2 \qquad \text{since } j^2 = -1$$

$$= (12-2)+j(8+3)$$

$$= 10+j11$$

EXAMPLE 2

Express the product of $2+j$, $-3+j2$,   and $1-j$ as a single complex number.

Then $(2+j)(-3+j2)(1-j) = (2+j)(-3+j2+j3-j^22)$

$$= (2+j)(-1+j5) \qquad \text{since } j^2 = -1$$

$$= -2-j+j10+j^25$$

$$= -7+j9 \qquad \text{since } j^2 = -1$$

Conjugate complex numbers

Consider: $(x+jy)(x-jy) = x^2+jxy-jxy-j^2y$

$$= x^2-(-1)y^2$$

$$= x^2+y^2$$

Hence we have two product of two complex numbers which produces a real number and therefore does not have a j term. If $x+jy$ represents a complex number then $x-jy$ is known as its *conjugate* (and vice versa). For example the conjugate of $(3+j4)$ is $(3-j4)$ and their product is

$$(3+j4)(3-j4) = 9+j12-j12-j^216 = 9-(-1)16$$

$$= 25 \text{ which is a real number.}$$

Division of complex numbers in algebraic form

Consider $\dfrac{(4+j5)}{(1-j\ )}$. We use the method of rationalising the denominator.

This means removing the j terms from the bottom line of the fraction. If we multiply $(1-j)$ by its conjugate $(1+j)$ the result will be a real number. Hence, in order not to alter the value of the given expression, we will multiply both the numerator and the denominator by $(1+j)$.

Thus
$$\frac{(4+j5)}{(1-j\ )} = \frac{(4+j5)(1+j)}{(1-j\ )(1+j)}$$

$$= \frac{4+j5+j4+j^2 5}{1-j\ +\ j-j^2}$$

$$= \frac{4+j9+(-1)5}{1-(-1)}$$

$$= \frac{-1+j9}{2}$$

$$= -\frac{1}{2}+j\frac{9}{2}$$

$$= -0.5+j4.5$$

EXAMPLE 3

The impedance Z of a circuit having a resistance and inductive reactance in series is given by the complex number $Z = 5+j6$.

Find the admittance Y of the circuit if $Y = \dfrac{1}{Z}$.

Now
$$Y = \frac{1}{Z} = \frac{1}{5+j6}$$

The conjugate of the denominator is $5-j6$ and therefore multiplying both the numerator and denominator by $5-j6$

then
$$Y = \frac{(5-j6)}{(5+j6)(5-j6)}$$

$$= \frac{5-j6}{25+j30-j30-j^2 36}$$

$$= \frac{5-j6}{25-(-1)36}$$

$$= \frac{5-j6}{61}$$

$$= \frac{5}{61}-j\frac{6}{61}$$

$$= 0.082-j0.098$$

EXAMPLE 4

Two impedances Z_1 and Z_2 are given by the complex numbers $Z_1 = 10+j15$ and $Z_2 = j70$. Find:

(a) the equivalent impedance, Z_1+Z_2, when the impedances are in series.

(b) the equivalent impedance, $\dfrac{1}{Z_1}+\dfrac{1}{Z_2}$, when the impedances are in parallel.

(a)
$$Z_1+Z_2 = (10+j15)+(j70)$$
$$= 10+j15+j70$$
$$= 10+j85$$

(b)
$$\frac{1}{Z_1}+\frac{1}{Z_2} = \frac{1}{(10+j15)}+\frac{1}{j70}$$

$$= \frac{10-j15}{(10+j15)(10-j15)}+\frac{j}{(j70)j}$$

$$= \frac{10-j15}{100+j150-j150-j^2225}+\frac{j}{j^270}$$

$$= \frac{10-j15}{100-(-1)225}+\frac{j}{(-1)70}$$

$$= \frac{10-j15}{325} - \frac{j}{70}$$

$$= \frac{10}{325}-j\frac{15}{325}-j\left(\frac{1}{70}\right)$$

$$= 0.0308-j0.0462-j0.0143$$

$$= 0.0308-j0.0605$$

Exercise 35

1) Add the following complex numbers:

(a) $3+j5$, $7+j3$, and $8+j2$
(b) $2-j7$, $3+j8$, and $-5-j2$
(c) $4-j2$, $7+j3$, $-5-j6$, and $2-j5$

2) Subtract the following complex numbers:

(a) $3+j5$ from $2+j8$ (b) $7-j6$ from $3-j9$
(c) $-3-j5$ from $7-j8$

3) Simplify the following expressions giving the answers in the form $x+jy$:

(a) $(3+j3)(2+j5)$ (b) $(2-j6)(3-j7)$
(c) $(4+j5)^2$ (d) $(5+j3)(5-j3)$
(e) $(-5-j2)(5+j2)$ (f) $(3-j5)(3-j3)(1+j)$

(g) $\dfrac{1}{2+j5}$ (h) $\dfrac{2+j5}{2-j5}$

(i) $\dfrac{-2-j3}{5-j2}$ (j) $\dfrac{7+j3}{8-j3}$

(k) $\dfrac{(1+j2)(2-j)}{(1+j)}$ (l) $\dfrac{4+j2}{(2+j)(1-j3)}$

4) Find the real and imaginary parts of:

(a) $1+\dfrac{j}{2}$ (b) $j3+\dfrac{2}{j^3}$ (c) $(j2)^2+3(j)^5-j(j)$

5) Solve the following equations giving the answers in the form $x+jy$:

(a) $x^2+2x+2=0$ (b) $x^2+9=0$

6) Three impedances Z_1, Z_2 and Z_3 are represented by the complex numbers $Z_1=2+j$, $Z_2=1+j$ and $Z_3=3+j2$.

(a) If they are connected in series find the equivalent impedance Z if $Z=Z_1+Z_2+Z_3$.
(b) If they are connected in parallel find the equivalent impedance Z if $Z=\dfrac{1}{Z_1}+\dfrac{1}{Z_2}+\dfrac{1}{Z_3}$.
(c) If Z_1 and Z_2 connected in parallel, are in series with Z_3 giving an equivalent impedance $Z=\dfrac{1}{Z_1}+\dfrac{1}{Z_2}+Z_3$, find Z.

7) Find the admittance Y of a circuit if $Y=\dfrac{1}{Z}$ where Z $=1.3+j0.6$.

THE ARGAND DIAGRAM

When plotting a graph, cartesian coordinates are generally used to plot the points. Thus the position of the point P (Fig. 18.1) is defined by the coordinates (3, 2) meaning that $x = 3$ and $y = 2$.

Complex numbers may be represented in a similar way on the Argand diagram. The real part of the complex number is plotted along the horizontal real-axis whilst the imaginary part is plotted along the vertical imaginary, or j axis.

However a complex number is denoted, not by a point but, as a phasor, a phasor being a line where regard is paid both to its magnitude and to its direction. Hence in Fig. 18.2 the complex number $4+j3$ is represented by the phasor $\overrightarrow{OQ}$, the end Q of the line being found by plotting 4 units along the real-axis and 3 units along the j-axis.

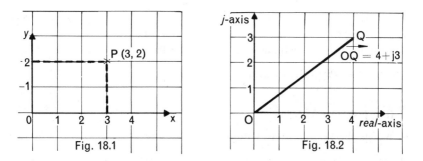

Fig. 18.1   Fig. 18.2

A single letter, the favourite being z, is often used to denote a phasor which represents a complex number. Thus if $z = x+jy$ it is understood that z represents a phasor and not a simple numerical value.

Four typical complex numbers z_1, z_2, z_3, and z_4 are shown on the Argand diagram in Fig. 18.3.

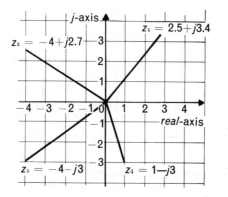

Fig. 18.3

A real number such as 2.7 may be regarded as a complex number with a zero imaginary part, i.e. $2.7 + j0$, and may be represented on the Argand diagram (Fig. 18.4) as the phasor $z = 2.7$ denoted by $\overrightarrow{OA}$ in the diagram.

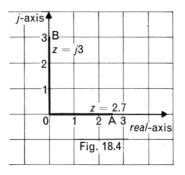

Fig. 18.4

A number such as $j3$ is said to be wholly imaginary and may be regarded as a complex number having a zero real part, i.e. $0 + j3$, and may be represented on the Argand diagram (Fig. 18.4) as the phasor $z = j3$ denoted by $\overrightarrow{OB}$ in the diagram.

The j-operator

Consider the real number 3 shown on the Argand diagram. in Fig 18.5

It may be regarded as a *phasor*, denoted by $\overrightarrow{OA}$, a phasor being a line where regard is paid both to its magnitude and to its direction. If we now multiply the real number 3 by j we obtain the complex number $j3$ which may be represented by the phasor $\overrightarrow{OB}$.

It follows that the effect of j on phasor $\overrightarrow{OA}$ is to make it become phasor $\overrightarrow{OB}$,

that is
$$\overrightarrow{OB} = j\overrightarrow{OA}$$

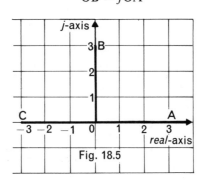

Fig. 18.5

Hence j is known as an operator (called the 'j-operator') which, when applied to a phasor, alters its direction by $90°$ in an anti-clockwise direction without changing its magnitude.

If we now operate on the vector $\overrightarrow{OB}$ we shall obtain, therefore, vector $\overrightarrow{OC}$.

In equation form this is, $\qquad \overrightarrow{OC} = j\overrightarrow{OB}$

but since $\overrightarrow{OB} = j\overrightarrow{OA}$ then $\qquad \overrightarrow{OC} = j(j\overrightarrow{OA})$

$$= j^2\overrightarrow{OA}$$

$$= -\overrightarrow{OA} \quad \text{since} \quad j^2 = -1.$$

This is true since it may be seen from the vector diagram that vector $\overrightarrow{OC}$ is equal in magnitude, but opposite in direction, to vector $\overrightarrow{OA}$.

Consider now the effect of the j-operator on the complex number $5+j3$.

In equation form this is: $j(5+j3) = j5+j(j3)$

$$= j5+j^23$$

$$= j5+(-1)3$$

$$= -3+j5$$

If phasor $z_1 = 5+j3$ and phasor $z_2 = -3+j5$, it may be seen from the Argand diagram in Fig. 18.6 that their magnitudes are the same but the effect of the operator j on z_1 has been to alter its direction by $90°$ anti-clockwise to give phasor z_2.

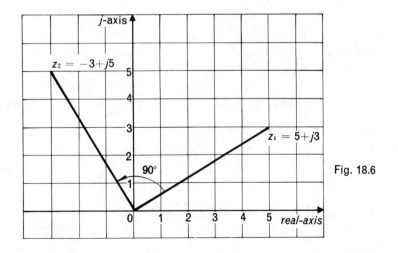

Fig. 18.6

Addition of Phasors

Consider the addition of the two complex numbers $2+j3$ and $4+j2$.

We have, $(2+j3)+(4+j2) = 2+j3+4+j2$

$$= (2+4)+j(3+2)$$

$$= 6+j5$$

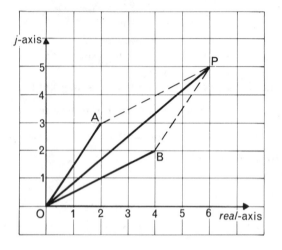

Fig. 18.7

On the Argand diagram shown in Fig. 18.7, the complex number $2+j3$ is represented by the phasor $\overrightarrow{OA}$, whilst $4+j2$ is represented by phasor $\overrightarrow{OB}$. The addition of the real parts is performed along the *real* axis and the addition of the imaginary parts is carried out on the *j*-axis.

Hence the complex number $6+j5$ is represented by the phasor $\overrightarrow{OP}$.

It follows that: $\overrightarrow{OP} = \overrightarrow{OB}+\overrightarrow{OA}$

$= \overrightarrow{OB}+\overrightarrow{BP}$ since $\overrightarrow{BP}$ is equal in magnitude and direction to $\overrightarrow{OA}$.

Hence the addition of phasors is similar to vector addition used when dealing with forces or velocities.

Subtraction of Phasors

Consider the difference of the two complex numbers, $4+j5$ and $1+j4$.

We have, $(4+j5)-(1+j4) = 4+j5-1-j4$

$$= (4-1)+j(5-4)$$

$$= 3+j$$

On the Argand diagram shown in Fig. 18.8, the complex number $4+j5$ is represented by the phasor $\overrightarrow{OC}$, whilst $1+j4$ is represented by the phasor $\overrightarrow{OD}$. The subtraction of the real parts is performed along the *real-axis*, and the subtraction of the imaginary parts is carried out along the *j*-axis. Now let $(4+j5)-(1+j4) = 3+j$ be represented by the phasor $\overrightarrow{OQ}$.

It follows that, $\overrightarrow{OQ} = \overrightarrow{OC}-\overrightarrow{OD}$

$$= \overrightarrow{OC}+\overrightarrow{CQ} \quad \text{since} \quad \overrightarrow{CQ} = -\overrightarrow{OD}.$$

As for phasor addition, the subtraction of phasors is similar to the subtraction of vectors.

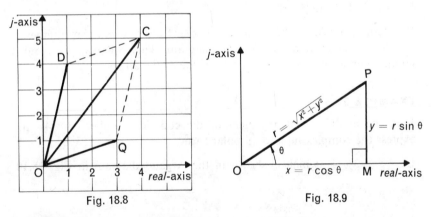

Fig. 18.8   Fig. 18.9

THE POLAR FORM OF A COMPLEX NUMBER

Let z denote the complex number represented by the phasor $\overrightarrow{OP}$ shown in Fig. 18.9. Then from the right angled triangle PMO we have:

$$z = x+jy$$

$$= r\cos\theta+j(r\sin\theta)$$

$$= r(\cos\theta+j\sin\theta)$$

The expression $r(\cos\theta + j\sin\theta)$ is known as *the polar form* of the complex number z. Using conventional notation it may be shown abbreviated as $r\underline{/\theta}$.

r is called the *modulus* of the complex number z and is denoted by $\bmod z$ or $|z|$.

Hence, from the diagram, $\qquad |z| = r = \sqrt{x^2 + y^2}$
using the theorem of Pythagoras for right-angled triangle PMO.

It should be noted that the plural of *modulus* is *moduli*.

The angle θ is called the *argument* (or amplitude) of the complex number z, and is denoted by $\arg z$ (or amp z).

Hence $\qquad\qquad\qquad \arg z = \theta$

and, from the diagram, $\qquad \tan\theta = \dfrac{y}{x}$

There are an infinite number of angles whose tangents are the same, and so it is necessary to define which value of θ to state when solving the equation $\tan\theta = \dfrac{y}{x}$. It is called the principal value of the angle and lies between $+180°$ and $-180°$.

We recommend that, when finding the polar form of a complex number, it should be sketched on an Argand diagram. This will help to avoid a common error of giving an incorrect value of the angle.

EXAMPLE 5

Find the modulus and argument of the complex number $3 + j4$ and express the complex number in polar form.

Let $z = 3 + j4$ which is shown in the Argand diagram in Fig. 18.10.

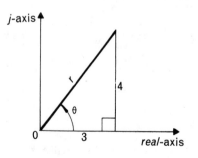

Fig. 18.10

Then $\qquad\qquad\qquad |z| = r = \sqrt{3^2+4^2} = 5$

and $\qquad\qquad\qquad \tan\theta = \dfrac{4}{3} = 1.3333$

$\therefore \qquad\qquad\qquad \theta = 53°\, 8'$

Hence in polar form: $\qquad z = 5(\cos 53°\, 8' + j\sin 53°\, 8')$

$\qquad\qquad\qquad\qquad\quad z = 5\underline{/53°\, 8'}$

EXAMPLE 6

Show the complex number $\ z = 3.5\ \underline{/-150°}\ $ on an Argand diagram, and find z in algebraic form.

Now z is represented by phasor $\overrightarrow{OP}$ in Fig. 18.11. It should be noted that since the angle is negative it is measured in a clockwise direction from the *real*-axis datum.

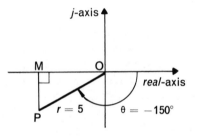

Fig. 18.11

In order to express z in algebraic form we need to find the lengths MO and MP. We use the right angled triangle PMO in which
$\text{P}\hat{\text{O}}\text{M} = 180° - 150° = 30°$

Now $\qquad\qquad \text{MO} = \text{PO}\cos \text{P}\hat{\text{O}}\text{M} = 5\cos 30° = 4.33$

and $\qquad\qquad \text{MP} = \text{PO}\sin \text{P}\hat{\text{O}}\text{M} = 5\sin 30° = 2.50$

Hence, in algebraic form, the complex number $\ z = -4.33 - j2.50.$

Multiplying numbers in polar form

Consider the complex number $\ z_1 = r_1\ \underline{/\theta_1} = r_1(\cos\theta_1 + j\sin\theta_1)$

and another complex number $\ z_2 = r_2\ \underline{/\theta_1} = r_2(\cos\theta_2 + j\sin\theta_2)$

Then the product of these two complex numbers:

$$z_1 \times z_2 = r_1(\cos\theta_1 + j\sin\theta_1) \times r_2(\cos\theta_2 + j\sin\theta_2)$$

$$= r_1r_2(\cos\theta_1 + j\sin\theta_1)(\cos\theta_2 + j\sin\theta_2)$$

$$= r_1r_2\{\cos\theta_1\cos\theta_2 + j\sin\theta_1\cos\theta_2$$
$$+ j\cos\theta_1\sin\theta_2 + j^2\sin\theta_1\sin\theta_2\}$$

$$= r_1r_2\{(\cos\theta_1\cos\theta_2 - \sin\theta_1\sin\theta_2) +$$
$$j(\sin\theta_1\cos\theta_2 + \cos\theta_1\sin\theta_2)\}$$

$$= r_1r_2\{\cos(\theta_1 + \theta_2) + j\sin(\theta_1 + \theta_2)\}$$

$$= r_1r_2 \underline{/(\theta_1 + \theta_2)}$$

Hence to multiply two complex numbers we multiply their moduli and add their arguments.

For example $6\underline{/17°} \times 3\underline{/35°} = 6 \times 3 \underline{/17° + 35°}$

$$= 18\underline{/52°}$$

Dividing numbers in polar form

It can be shown that the division of two complex numbers, using a method similar to that for finding the product of two complex numbers, is given by:

$$\frac{z_1}{z_2} = \frac{r_1\underline{/\theta_1}}{r_2\underline{/\theta_2}} = \frac{r_1}{r_2}\underline{/\theta_1 - \theta_2}$$

Hence we divide two complex numbers we divide their moduli and subtract their arguments.

For example: $\dfrac{5\underline{/33°\ 55'}}{3\underline{/-23°\ 40'}} = \dfrac{5}{3}\underline{/(33°\ 55') - (-23°\ 40')}$

$$= 1.67\underline{/33°\ 55' + 23°\ 40'}$$

$$= 1.67\underline{/57°\ 35'}$$

EXAMPLE 7

A simple circuit which has a resistance R in series with an inductive reactance X_L has an impedance Z is given by the complex number

$$Z = R + jX_L$$

A simple circuit which has a resistance R in series with a capacitive reactance X_C has an impedance Z given by the complex number

$$Z = R - jX_C.$$

Using the above relationships find the resistance and the inductive or capacitive reactance for each of the following impedances:

(a) $8+j12$ (b) $20-j80$ (c) $40\;\underline{/25°}$ (d) $100\;\underline{/-20°}$

(a) Here $Z = 8+j12$, and since it is of the form $Z = R+jX_L$ we can say that the: resistance $R = 8$
and the inductive reactance $X_L = 12$

(b) Here $Z = 20-j80$, and since it is of the form $Z = R-jX_C$ we can say that the: resistance $R = 20$
and the capacitive reactance $X_C = 80$

(c) The complex number $Z = 40\;\underline{/25°}$ is shown on the Argand diagram in Fig. 18.12. If we let $Z = x+jy$, then from the diagram:

$$x = 40\cos 25° \quad \text{and} \quad y = 40\sin 25°$$

$$= 36.3 \qquad\qquad\qquad = 16.9$$

Hence $Z = 36.3+j16.9$ which is of the form $Z = R+jX_L$ and we can say that resistance $R = 36.3$
and the inductive reactance $X_L = 16.9$

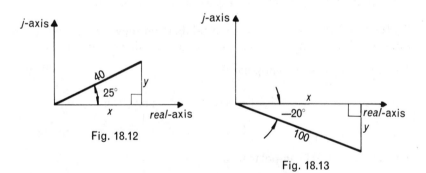

Fig. 18.12

Fig. 18.13

(d) The complex number $Z = 100\;\underline{/-20°}$ is shown on the Argand diagram in Fig. 18.13. If we let $Z = x+jy$, then from the diagram

$$x = 100\cos 20° \quad \text{and} \quad y = 100\sin 20°$$

$$= 94.0 \qquad\qquad\qquad = 34.2$$

but we can see from the diagram that the y value is negative hence $Z = 94.0-j34.2$, which is of the form $Z = R-jX_C$ and we can say that the resistance $R = 94.0$
and the capacitive reactance $X_C = 34.2$.

EXAMPLE 8

The potential difference across a circuit is given by the complex number $V = 50+j30$ volts, and the current is given by the complex number $I = 9+j4$ amperes.   Find:

(a) the phase difference (i.e. the angle ϕ in Fig. 18.14) between the phasors for V and I
(b) the power, given that power $= |V| \times |I| \times \cos \phi$ watts.

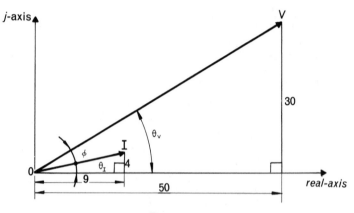

Fig. 18.14

Fig. 18.14 shows a sketch of the Argand diagram showing the phasors for I and V.   Phasors in electrical work are usually shown with arrows.

To find $V = 50+j30$ in polar form:

$$|V| = \sqrt{50^2+30^2} \quad \text{and} \quad \tan \theta_V = \frac{30}{50}$$

$$= 58.3 \qquad \therefore \qquad \theta_V = 30° \, 58'$$

To find $I = 9+j4$ in polar form:

$$|I| = \sqrt{9^2+4^2} \quad \text{and} \quad \tan \theta_I = \frac{4}{9}$$

$$= 9.8 \qquad \therefore \qquad \theta_I = 23° \, 58'$$

(a) The phase difference $\phi = \theta_V - \theta_I$

$$= 30° \, 58' - 23° \, 58'$$

$$= 7°$$

(b)   power $= |V| \times |I| \times \cos \phi$

$$= 58.3 \times 9.8 \times \cos 7°$$

$$= 567 \text{ watts}$$

Exercise 36

1) Show, indicating each one clearly, the following complex numbers on a single Argand diagram: $4+j3$, $-2+j$, $3-j4$, $-3.5-j2$, $j3$ and $-j4$.

2) Find the moduli and arguments of the complex numbers $3+j4$ and $4-j3$.

3) If the complex number $z_1 = -3+j2$ find $|z_1|$ and $\arg z_1$.

4) If the complex number $z_2 = -4-j2$ find $|z_2|$ and $\arg z_2$.

5) Express each of the following complex numbers in polar form:

(a) $4+j3$ (b) $3-j4$ (c) $-3+j3$ (d) $-2-j$ (e) $j4$
(f) $-j3.5$.

6) Convert the following complex numbers, which are given in polar form, into Cartestian form:

(a) $3\underline{/45°}$ (b) $5\underline{/154°}$ (c) $4.6\underline{/-20°}$ (d) $3.2\underline{/-120°}$

7) Simplify the following products of two complex numbers, given in polar form, expressing the answer in polar form:

(a) $8\underline{/30°} \times 7\underline{/40°}$ (b) $2\underline{/-20°} \times 5\underline{/-30°}$ (c) $5\underline{/120°} \times 3\underline{/-30°}$
(d) $7\underline{/-50°} \times 3\underline{/-40°}$

8) Simplify the following divisions of two complex numbers, given in polar form, expressing the answer in polar form:

(a) $\dfrac{8\underline{/20°}}{3\underline{/50°}}$ (b) $\dfrac{10\underline{/-40°}}{5\underline{/20°}}$ (c) $\dfrac{3\underline{/-15°}}{5\underline{/-6°}}$ (d) $\dfrac{1.7\underline{/35°\ 17'}}{0.6\underline{/-9°\ 22'}}$

9) Three complex numbers z_1, z_2 and z_3 are given in polar form by $z_1 = 3\underline{/35°}$, $z_2 = 5\underline{/28°}$ and $z_3 = 2\underline{/-50°}$. Simplify:

(a) $z_1 \times z_2 \times z_3$ giving the answer in polar form.

(b) $\dfrac{z_1 \times z_2}{z_3}$ giving the answer in algebraic form.

10) If the complex number $z = 2-j3$ express in polar form:

(a) $\dfrac{1}{z}$ (b) z^2

11) The admittance Y of a circuit is given by $Y = \dfrac{1}{Z}$.

(a) If $Z = 3+j5$ find Y in polar form.
(b) If $Z = 17.4\underline{/42°}$ find Y in algebraic form.

12) Using the notation and information given in the data for the worked Example 7 of the text (p. 250), find the resistance and the inductive or capacitive reactance for each of the following impedances:

(a) $4.5+2.2j$ (b) $23-35j$ (c) $29.6\,\underline{/23°\,22'}$ (d) $7\,\underline{/-12°}$

13) The potential difference across a circuit is given by the complex number $V = 40+j35$ volts and the current is given by the complex number $I = 6+j3$ amperes. Sketch the appropriate phasors on an Argand diagram and find:

(a) the phase difference (i.e. the angle ϕ) between the phasors for V and I,
(b) the power, given that power $= |V|\times|I|\times\cos\phi$.

SUMMARY

$$j = \sqrt{-1} \quad \text{or} \quad j^2 = -1$$

The algebraic form of a complex number is $x + jy$.

The polar form of a complex number is $r(\cos\theta+j\sin\theta)$ or $r\,\underline{/\theta}$.

The conjugate of $(x+jy)$ is $(x-jy)$.

The conjugate of $(x-jy)$ is $(x+jy)$.

If a complex number is multiplied by its conjugate the result is a real number.

The effect of j as an operator on a phasor representing a complex number is to change its direction by 90° anti-clockwise, without any alteration in its magnitude.

If a complex number z in its algebraic form is $z = x+jy$,

and in its polar form is $z = r(\cos\theta+j\sin\theta)$ or $r\,\underline{/\theta}$

then $\text{mod } z = |z| = r = x^2+y^2$

and $\arg z = \theta$ where $\tan\theta = \dfrac{y}{x}$

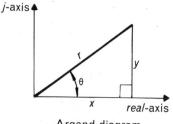

Argand diagram

To multiply two complex numbers in polar form, we multiply their moduli and add their arguments. Thus $(r_1 \underline{/\theta_1}) \times (r_2 \underline{/\theta_2}) = (r_1 \times r_2) \underline{/\theta_1 + \theta_2}$

To divide two complex numbers in polar form, we divide their moduli and subtract their arguments. Thus: $\dfrac{r_1 \underline{/\theta_1}}{r_2 \underline{/\theta_2}} = \left(\dfrac{r_1}{r_2}\right) \underline{/\theta_1 - \theta_2}$

ANSWERS

Exercise 1

1) $15x^2+14x-1$ **2)** $3.5t^{-0.5}-1.8t^{-0.7}$
3) 1.25   **4)** 1   **5)** $5, -3$

Exercise 2

1) $6(3x+1)$ **2)** $-15(2-5x)^2$
3) $-2(1-4x)^{-\frac{1}{2}}$ **4)** $-7.5(2-5x)^{\frac{1}{2}}$
5) $-4(4x^2+3)^{-2}$ **6)** $3\cos(3x+4)$
7) $5\sin(2-5x)$ **8)** $8\sin 4x \cos 4x$

9) $21(\sin 7x)/\cos^4 7x$ **10)** $2\cos\left(2x+\dfrac{\pi}{2}\right)$

11) $-3(\sin x)\cos^2 x$ **12)** $-\dfrac{\cos x}{\sin^2 x}$

13) $\dfrac{1}{x}$ **14)** $-\dfrac{9}{x}$

15) $\dfrac{1}{2(2x-7)}$ **16)** $-\dfrac{1}{e^x}$

17) $6e^{(3x+4)}$ **18)** $8e^{(8x-2)}$

19) $\dfrac{2}{3}(1-2t)^{-4/3}$ **20)** $\dfrac{3}{4}\cos(\tfrac{3}{4}\theta-\pi)$

21) $\{-\sin(\pi-\varphi)\}/\cos^2(\pi-\varphi)$

22) $-\dfrac{1}{2x}$ **23)** $Bk e^{(kt-b)}$

24) $-\dfrac{1}{3}e^{(1-x)/3}$

Exercise 3

1) (a) $\sin x + x\cos x$
  (b) $e^x(\tan x + \sec^2 x)$
  (c) $1+\log_e x$
2) $\cos^2 t - \sin^2 t$
3) $2(\tan\theta)\cos 2\theta + (\sec^2\theta)\sin 2\theta$
4) $e^{4m}(4\cos 3m - 3\sin 3m)$
5) $3x(1+2\log_e x)$ **6)** $6e^{3t}(3t^2+2t-3)$
7) $1-3z+(1-6z)\log_e z$

8) (a) $\dfrac{1}{(1-x)^2}$ (b) $\dfrac{1-2\log_e x}{x^3}$

  (c) $\dfrac{e^x(\sin 2x - 2\cos 2x)}{\sin^2 2x}$

9) $\dfrac{11}{(3-4z)^2}$

10) $-\dfrac{2(\sin 2t + \cos 2t)}{e^{2t}}$

11) $-\text{cosec}^2\theta$

Exercise 4

1) 21.59   **2)** -3.77 **3)** 1.005
4) -0.3573 **5)** 0   **6)** 26.15
7) 5.882   **8)** -4.050 **9)** 0.5
10) 0.5266

Exercise 5

1) $18x$, 54   **2)** $-187, 254$ **3)** 29.3
4) $-\dfrac{1}{x^2}, -0.277$ **5)** 15, 0
6) -89.0 **7)** -0.115 **8)** 7.52
9) $-4.6, -21.2$ **10)** 222   **11)** 2.94

Exercise 6

1) 42 m/s   **2)** 6 m/s^2
3) (a) 6 m/s   (b) 2.41 or -0.41 s
  (c) 6 m/s^2 (d) 1 s
4) -0.074 m/s, 0.074 m/s^2
5) 10 m/s, 30 m/s   **6)** 3.46 m/s
7) (a) 4 rad/s   (b) 36 rad/s^2
  (c) 0 s or 1 s
8) (a) -2.97 rad/s (b) 0.280 s
  (c) -8.98 rad/s^2 (d) 1.57 s
9) 62.5 kJ

Exercise 7

1) (a) 11 (max), -16 (min)
  (b) 4 (max), 0 (min)
  (c) 0 (max), -32 (min)
2) (a) 54   (b) $x=2.5$ (c) $x=-2$
3) $(3, -15), (-1, 17)$
4) (a) -2 (b) 1   (c) 9
5) (a) 12   (b) 12.48
6) 15 mm   **7)** 10 m   **8)** 4
9) 108 000 mm^3
10) radius = height = 4.57 m
11) 405 mm
12) diameter = 28.9 mm, height 14.4 mm
13) 5.76 m   **14)** 2.15   **15)** 84.8 mm
16) $\cos^2 a$

Exercise 8

1) $\dfrac{5}{4}x^4+x^2+\dfrac{4}{x}+k$ 2) $\dfrac{2}{3}x^{3/2}+2x^{\frac{1}{2}}+k$

3) 0.841   4) 1   5) 0.086
6) 6.11   7) 0   8) 3.63
9) 4.67   10) 0

Exercise 9

1) 14.7   2) 2.67   3) -0.167
4) 4.5   5) $+0.25, -0.25$
7) 2.67   8) 2.19   10) 0.866
11) 7.25   12) 3.69   13) 1.26

Exercise 10

1) 57.4   2) 402   3) 0.0761
4) 171   5) 16.8   6) 0.105
7) (a) 262 000 mm³   (b) 58 700 mm³
8) 14.6 litres 9) 230 000 mm³

Exercise 11

1) 251 000 mm³   2) 24.3 kg
3) 189 000 mm³   4) 1.91 kg
5) 20 500 mm³   6) 393 000 mm³
7) 56 300 mm³   8) 3.37 kg
9) 146 100 mm³   10) 87 190 mm³

Exercise 12

1) 3970 mm²   2) 158 000 mm²
3) 3680 mm²   4) 160 000 mm²
5) 66 000 mm²   6) 6150 mm²

Exercise 13

1) $\bar{x} = 38.7$ mm, $\bar{y} = 41.8$ mm
2) $\bar{x} = 68.3$ mm, $\bar{y} = 73.8$ mm
3) $\bar{x} = 19.1$ mm, $\bar{y} = 28.3$ mm
4) $\bar{x} = 50.0$ mm, $\bar{y} = 44.6$ mm
5) $\bar{y} = 42.3$ mm
6) $\bar{x} = 86.7$ mm, $\bar{y} = 60.7$ mm
7) $\bar{y} = 29.7$ mm   8) $\bar{y} = 10.6$ mm

Exercise 14

1) $I_{XX} = 856$ cm⁴, $I_{YY} = 1184$ cm⁴
2) $I_{XX} = 1410$ cm⁴, $I_{YY} = 354$ cm⁴
3) $I_{XX} = 296$ cm⁴, $I_{YY} = 133$ cm⁴
4) $I_{XX} = 521$ cm⁴, $I_{YY} = 1100$ cm⁴
5) $I_{XX} = 492$ cm⁴, $I_{YY} = 172$ cm⁴
6) $I_{XX} = 136$ cm⁴, $I_{YY} = 60$ cm⁴

Exercise 15

1) $I_{XX} = 375$ cm⁴, $I_{YY} = 266$ cm⁴
2) $I_{XX} = 131$ cm⁴, $I_{YY} = 227$ cm⁴
3) $I_{XX} = 284$ cm⁴, $I_{YY} = 227$ cm⁴
4) $I_{XX} = I_{YY} = 168$ cm⁴
5) 1410 cm⁴, 296 cm⁴, 521 cm⁴, 492 cm⁴
6) (a) 258 cm⁴   (b) 517 cm⁴
   (c) 2520 cm⁴
7) (a) 1600 cm⁴   (b) 1820 cm⁴
   (c) 909 cm⁴

Exercise 16

1) $y = \frac{1}{2}x^2+1$ 2) $y = x^3-128$
3) $y = \dfrac{1}{3}x^3-\dfrac{5}{2}x^2+k,\ y = \dfrac{1}{3}x^3-\dfrac{5}{2}x^2+18.7$
4) $y = 1.5x^4+1.67x^3+7x-49.4$
5) $y = x^2$

Exercise 17

1) $y = Ae^{3x},\ y = 2e^{3x},\ 807$
2) $s = 12.25e^{-0.560t},\ 3.24$
3) $I = 10e^{-33.3t},\ 5.14$ A
4) 30 years
5) $Q = 0.0015e^{-41.7t},\ 0.000\ 652$ farad,
   0.009 72 seconds

Exercise 18

1) 9.17 m, 24° 12′   2) 257.3 mm, 97° 2′
3) 111.5 mm, 77° 6′   4) 26.17 mm
5) 144.5 mm   6) 4.78 m, 4.09 m
7) 13.2 m²   8) 93.3 m²
9) 4105 mm²   10) 576 mm²

Exercise 19

1) (a) 42.72 mm   (b) 854.4 mm²
   (c) 54° 11′
2) (a) 78.5 mm   (b) 20.3 mm
   (c) 12° 14′
3) (a) 34° 55′   (b) 31° 18′
4) 180 000 mm²
5) (a) 3.37 m   (b) 25° 31′
   (c) 7.83 m
6) 86.0 mm, 94.9 mm, 24° 56′
7) (a) 50° 46′   (b) 104° 30′
8) 28° 44′

Exercise 21

4) 4, 2.094 seconds   5) 3 sin 2t

Exercise 22

4) (a) $\sin\left(\omega t + \dfrac{\pi}{2}\right)$ (c) $\sin \omega t$

(b) $\sin(\omega t - \pi)$ (d) $\sin\left(\omega t - \dfrac{\pi}{6}\right)$

Exercise 23

1) (a) $-\sin x$ (b) $-\cos x$
(c) $\cos x$ (d) $\sin x$
2) $2\sin\theta\cos\theta$
4) (a) $\dfrac{16}{65} = 0.2462$ (b) $\dfrac{33}{65} = 0.5077$
6) $46°\ 49'$ **7)** 0.2588
8) 0 **9)** $\theta = 10°\ 53'$
10) (a) 0.5505 (b) -0.5505

Exercise 24

1) $3.606\sin(\theta + 33\ 41')$
2) $7.071\sin(t + 1.429)$
3) $8.602\sin(t - 0.951)$
4) 8.602, 2.521 radians
5) $223.6\sin(300t + 0.464)$, 223.6
6) (a) $9.33\sin x + 11.16\cos x$
(b) $14.55\sin(x + 50°\ 6')$

Exercise 25

1) $1 + 5z + 10z^2 + 10z^3 + 5z^4 + z^5$
2) $p^6 + 6p^5q + 15p^4q^2 + 20p^3q^3 + 15p^2q^4 + pq^5 + q^6$.
3) $x^4 - 12x^3y + 54x^2y^2 - 108xy^3 + 81y^4$
4) $32p^5 - 80p^4q + 80p^3q^2 - 40p^2q^3 + 10pq^4 - q^5$
5) $128x^7 + 448x^6y + 672x^5y^2 + 560x^4y^3 + 280x^3y^4 + 84x^2y^5 + 14xy^6 + y^7$
6) $x^3 + 3x + \dfrac{3}{x} + \dfrac{1}{x^3}$
7) $1 + 12x + 66x^2 + 220x^3$
8) $1 - 28x + 364x^2 - 2912x^3$
9) $p^{16} + 16p^{15}q + 120p^{14}q^2 + 560p^{13}q^3$
10) $1 + 30y + 405y^2 + 3240y^3$
11) $x^{18} - 27x^{16}y + 324x^{14}y^2 - 2268x^{12}y^3$
12) $x^{22} + 11x^{18} + 55x^{14} + 165x^{10}$
14) 5.6% too large   **15)** 10.2% too small
16) 2.9% decrease   **17)** 20.5% increase

Exercise 26

1) 3   **2)** 3   **3)** 4
4) 3   **5)** 9   **6)** 64
7) 100   **8)** 1   **9)** 2
10) 3

Exercise 27

1) (a) 2.212   (b) 1.795   (c) 3.373
(d) 5.698   (e) -0.525 (f) -4.942
2) 0.00319   **3)** 0.0146
4) (a) 1.35   (b) 12.18   (c) 90.02
(d) 0.67   (e) 0.37   (f) 0.03
5) $y = 1.82$, $x = 0.84$
6) $x = -1.39$, $y = -0.5$
7) (a) $T = 631$ (b) $s = 1.12$
8) 4.34 mA per second
9) (a) 0.18 seconds
(b) 1200 V per second
10) 10 years

Exercise 28

1) $a = 94$, $b = 0.35$ **2)** $E = 1.5$, $r = 0.6$
3) $a = 3$, $n = 5$
4) $n = 4$, for $V = 80$ read $V = 70$
5) $t = 0.3m^{1.5}$
6) $k = 100$, $n = -1.2$
7) $a = 245$, $b = 33$ **8)** $\theta = 0.5$, $k = 5$
9) $I = 0.02$, $T = 0.2$

Exercise 29

1) $\frac{1}{13}$ **2)** $\frac{1}{9}$ **3)** $\frac{5}{13}$
4) (a) $\frac{2}{5}$ (b) $\frac{1}{4}$ (c) $\frac{3}{5}$
(d) $\frac{17}{20}$ (e) $\frac{1}{5}$ **5)** $\frac{1}{2}$
6) (a) 2% (b) 0.02   **7)** 0.9

Exercise 30

1) $\frac{3}{13}$ **2)** 0.45
3) 0.5   **4)** $\frac{1}{3}$
5) $\frac{1}{36}$
6) (a) $\frac{1}{4}$ (b) $\frac{1}{8}$
(c) $\frac{7}{8}$ (d) $\frac{1}{2}$
7) (a) (i) $\frac{1}{8}$ (ii) $\frac{3}{8}$
(b) (i) $\frac{1}{24}$ (ii) $\frac{1}{12}$
8) (a) $\frac{8}{27}$ (b) $\frac{1}{27}$
(c) $\frac{2}{9}$ (d) $\frac{4}{9}$
9) (a) $\frac{4}{27}$ (b) $\frac{5}{27}$
10) (a) $\frac{1}{3}$ (b) $\frac{5}{12}$
(c) $\frac{1}{24}$ (d) $\frac{1}{12}$
11) (a) 0.32   (b) 0.03
(c) (i) 0.0196   (ii) 0.04
(iii) 0.1652   (iv) 0.1472
12) (a) $\frac{15}{92}$ (b) $\frac{91}{276}$
(c) $\frac{35}{138}$ (d) $\frac{35}{138}$

Exercise 31

1) 0.6561, 0.2916, 0.0486, 0.0036, 0.0001

2) (a) 0.3585   (b) 0.2641
3) 0.1916
4) (a) 0.3632 (b) 0.3725 (c) 0.2643
5) 0.1108   **6)** 0.3232

Exercise 32

1) 0.6703, 0.2681, 0.0536, 0.0072, 0.0007
2) 0.2240
3) (a) 0.2707   (b) 0.5940
4) 0.4422   **5)** 56.65%
6) (a) 0.0498   (b) 0.1992   (c) 0.5768
7) 0.7769   **8)** 0.4060

Exercise 33

1) (a) −1.6   (b) −0.4   (c) 1.0
    (d) 2.2
2) (a) 0.3849   (b) 0.2734   (c) 0.4865
    (d) 0.0487   (e) 0.2730   (f) 0.2148
    (g) 0.0174   (h) 0.8051   (i) 0.8686
    (j) 0.9936
3) (a) 18.60 ± 0.06 mm
    (b) (i) 2.3% (ii) 0.6%   (iii) 53.3%
4) (a) 28.6%   (b) 66.8%   (c) 88.8%
    (d) 2.3%   (e) 6.7%
5) 1810
6) $\bar{x} = 10.000$, $\sigma = 0.0128$
    (a) 2.5%   (b) 0.3%   (c) 75.8%
7) $\bar{x} = 1.6847$ m, $\sigma = 0.0956$ m, 31.8%
8) $\bar{x} = 170.025$, $\sigma = 0.765$ kg, 89.8%
9) $\bar{x} = 12.50$, $\sigma = 0.0112$, 598
10) 115, 673   **11)** 95
12) $\bar{x} = 20.00$ μF, $\sigma = 0.0129$ μF
13) $\bar{x} = 170.025$ kg, $\sigma = 0.765$ kg
14) 67.4%   **15)** 109
16) 6.5%   **17)** 48.0%

Exercise 34

1) 0.0133 kg m²   **2)** 104 mm
3) 21.4 kg m²
4) (a) 8390 kg m²   (b) 32 300 kg m²

5) (a) 288 kg m²   (b) 1050 kg m²
6) 17 700 kg m²   **7)** 5.83 kg m²

Exercise 35

1) (a) $18+j10$ (b) $-j$ (c) $8-j10$
2) (a) $-1+j3$ (b) $-4-j3$ (c) $10-j3$
3) (a) $-9+j21$ (b) $-36-j32$
    (c) $-9+j40$ (d) 34 (e) $-21-j20$
    (f) $18-j30$ (g) $0.069-j0.172$
    (h) $-0.724+j0.690$
    (i) $-0.138-j0.655$
    (j) $0.644+j0.616$ (k) $3.5-j0.5$
    (l) $0.2+j0.6$.
4) (a) $1, j0.5$ (b) $0, j5$ (c) $-3, j3$
5) (a) $-1\pm j$ (b) $\pm j3$
6) (a) $6+j4$ (b) $1.131-j0.854$
    (c) $3.9+j1.3$
7) $0.634-j0.293$

Exercise 36

2) Mod 5, Arg 53° 8′; Mod 5, Arg −36°52
3) 3.61, 146° 19′
4) 4.47, −153° 26′
5) (a) 5 $\underline{/36°\ 52'}$ (b) 5 $\underline{/-53°\ 8'}$
    (c) 4.24 $\underline{/135°}$
    (d) 2.24 $\underline{/-153°\ 26'}$
    (e) 4 $\underline{/90°}$ (f) 3.5 $\underline{/-90°}$
6) (a) $2.12+j2.12$ (b) $-4.49+j2.19$
    (c) $4.32-j1.57$ (d) $-1.60-j2.77$
7) (a) 56 $\underline{/70°}$ (b) 10 $\underline{/-50°}$
    (c) 15 $\underline{/90°}$ (d) 21 $\underline{/-90°}$
8) (a) 2.67 $\underline{/-30°}$ (b) 2 $\underline{/-60°}$
    (c) 0.6 $\underline{/-9°}$ (d) 2.83 $\underline{/44°\ 39'}$
9) (a) 30 $\underline{/33°}$ (b) $-2.93+j6.90$
10) (a) 0.277 $\underline{/56°\ 19'}$
    (b) 13 $\underline{/-112°\ 38'}$
11) (a) 0.172 $\underline{/-59°\ 2'}$
    (b) $0.0427-j0.0385$
12) (a) $r = 4.5$, $X_L = 2.2$
    (b) $R = 23$, $X_C = 35$
    (c) $R = 27.2$, $X_L = 11.7$
    (d) $R = 6.85$, $X_C = 1.46$
13) (a) 14° 37′   (b) 345 watts.

INDEX

Addition law of probability 188
Amplitude 132
Angle between a line and a plane 125
    two planes 126
Angular and time scales 133
Areas by integration 46
    under normal curve 215
Argand diagram 243
Arithmetic mean 211

Binomial coefficients 196
    distribution 199
    expansion 152

Centroids of areas 70
Complex numbers 236
Compound angle formulae 145
Conjugate complex numbers 239
Cosine rule 119
Cycle 134

Differentiation 1
    of trigonometrical functions 6
    of exponential functions 12
    of logarithmic functions 10

Empirical probablity 186
Exponential functions 158
    differentiation of 12
Exponential graphs 162

Families of curves 109
Frequency 135
Function of a function 4

Graphs of trigonometrical functions 136

Integration 44

j-operator 244

Law of a straight line 171
Logarithmic scales 173
    graph paper 176
Logarithms, theory of 159

Maxima and minima 33
Mean 211
Moments of inertia 228
Multiplication law of probablity 190

Normal distribution 210

Pappus' theorem 98
Parallel axis theorem 94, 232
Peak value 132
Period 134
Perpendicular axis theorem 102
Phase angle 140
Poisson probability 202
Polar form of complex numbers 247
Probability 184
    scale 185
Polar second moment of area 100
Product, differentiation of 15

Quotient, differentiation of 17

Radians and degrees 135
Radius of gyration 105
Repeated trials 194

Second derivative 24
Second moment of area 84
Sine rule 119
Small errors 154
Solution of triangles 119
Standard deviation 212
Surface areas of solids of revolution 66

Three dimensional problems 123
Time base 135
Total probability 187
Trigonometrical functions, differentiation
    of 6
Turning points 33

Units of moment of inertia 230
    second moments of area 86

Velocity and acceleration 28
Volumes by integration 52
Volumes of solids of revolution 58